CHINA'S GREAT LIBERAL
OF THE 20TH CENTURY –

HU SHIH

A PIONEER OF
MODERN CHINESE LANGUAGE

By Mark O'Neill

Hu Shih in March 1927

INTRODUCTION

When you read the story of Hu Shih, you feel great optimism for the future of China. He was a man with a wealth of knowledge who wrote 44 books on many subjects, including three in English. A child of the Qing dynasty and the Western enlightenment, he had a mastery of English and Chinese and was equally at home in both worlds. He had every reason to be proud and conceited, but he was not. He was approachable and humorous and he made friends everywhere.

At his home in Beijing in the 1930s, he held open house on Sunday mornings – beggars and petitioners, students and passers-by, all were welcome. Among his many achievements were: being a professor at Beijing University and becoming its president in 1946; serving as his country's ambassador in Washington during the critical years of 1938-1942; playing a key role in bringing the U.S. into the war on China's side. Hu was President of Academia Sinica for the last four years of his life. He edited and wrote for magazines that expressed the new, progressive ideas that flooded into China after the end of the last dynasty in 1911. His life, books and ideas influenced and changed thousands of his countrymen. He was China's greatest public intellectual of the 20th century.

December 2021 marks the 130th anniversary of Hu's birth. Joint Publishing and I decided that we should offer a biography to mark this important date. While most Chinese know of him, they are ignorant of the details of his rich life. In school textbooks in the Chinese world, he is not a major figure. Other than Sinologists, foreigners have never heard of him. "Who is Hu?" was the response when I asked my foreign friends about him.

Thanks to his prolific writing on many subjects, Chinese scholars have written comprehensive studies of his life and work. I express here my deep admiration and respect for them. Those studies are the main source of material for this book. I am no more than a humble student who has had the good fortune to read and learn from them. We list their names in the Thanks and Acknowledgements section. We thank them very much for their meticulous scholarship and deep insights.

Hu was born in Shanghai in December 1891 into a traditional family with roots in rural Anhui province in East China. After secondary education in Shanghai, he spent seven years at Cornell and Columbia, two of the top universities in the United States. The experience opened his eyes to the advanced industrial world and formed his character – sociable, optimistic and fluent in English, making him at ease with both foreigners and Chinese.

Hu was fortunate in the era in which he lived. His professional life ran from 1917, when he returned from the U.S., until his passing in 1962. The 38 years between the Xinhai Revolution of 1911 and the foundation of the People's Republic in 1949 were the most liberal in China's history. Intellectuals had the opportunity to express and publish their opinions, on politics, society, culture, literature and religion, in a way that was impossible before and after. Hu took full advantage of this freedom. Even as a student in the U.S., he wrote articles on major issues; they were published in magazines in China and reached a wide audience.

The second factor working in his favour was the fall of the Qing dynasty in 1911 and the founding of Asia's first republic. The end of the imperial system unleashed a ferment of ideas and debate. How should China build a new society and form of government? Which elements of traditional society should Chinese retain and which should they discard? This created an ideal stage for Hu's prolific and restless mind. Everyone wanted to hear his ideas on reforming marriage, improving the status of women, the written language, Confucianism, science and democracy.

He also had the good fortune to spend seven years in the country that was overtaking Britain and Germany as the world's number one industrial power and source of innovation. It was a textbook of ideas and experiments on how to drive an economy and remodel society. Few foreign students in the United States used their time there as fruitfully as he. In addition to his academic work, he threw himself into learning about American society, making a wide network of friends, among fellow Chinese and Americans. With one of them, Edith Clifford Williams, he developed a friendship that lasted 50 years.

He became an accomplished public speaker in English – in fact, he had so many invitations that, in 1915, the philosophy department of Cornell

University withdrew his fellowship. His professors said he was spending too much time on public speaking and not enough on studying Immanuel Kant and Georg Hegel! He acquired a broader knowledge of American politics and society than many Americans. It was this Western education – then rare among Chinese – that earned him a professorship at Beijing University at the tender age of 26.

Hu was also blessed by the kind of Americans he met – well educated, enlightened and favourable to foreign students. An Exclusion Act of 1882 had banned immigration of Chinese labourers; it was the result of racism and resentment among whites against those willing to work for lower wages. But the American professors, their wives and students whom Hu encountered in Cornell and Columbia welcomed him into their lives, academic and social.

His countrymen in New York and San Francisco lived in the ghettos of Chinatown, fearful of violence and discrimination. But Hu was invited into the dining rooms and country picnics of the American elite and their families and treated as an equal. Some became his lifelong friends. It was this experience that made him so favourable to American thinking and practices. He researched which ones he could take home and implement in China.

These seven years also gave him personal skills and networks that would prove invaluable throughout his life. He was able to raise foreign money for Chinese universities and railways, becoming a popular speaker at universities in North America and Britain. He became the most famous Chinese in the U.S. And, as ambassador in Washington, he played a critical role in the autumn of 1941 – he stopped an agreement between Japan and the U.S. that would have prevented the attack on Pearl Harbour. Without that attack, the U.S. may not have entered World War Two and China not won the war.

His greatest legacy was the introduction of the vernacular as the standard for written Chinese, replacing the classical form. He set out his case in an article – written in a New York apartment – that appeared in a Beijing magazine *New Youth* in January 1917. It had a nationwide impact. Within five years, the Ministry of Education was preparing textbooks in the vernacular; newspapers and magazines switched to it, as did many authors. The change was faster than anyone could have imagined. This was a revolution, giving millions of ordinary people access to the written word – and the knowledge it imparted – they had not before.

Promoting the vernacular was one of many causes Hu took up, in speeches, newspapers and magazines. He advocated the end of foot binding for women, the right of young people to choose their own marriage partners, birth control and sexual equality. He worked for an education that encouraged students to be critical and think for themselves and a mind-set that was sceptical and only accepted things that were proved.

To catch up with the West, China must learn science and technology; it must not believe that its 'spiritual culture' was superior to that of the 'material' West. "What spirituality is there in a civilization which tolerated so cruel and inhuman an institution as foot-binding for women for over 1,000 years?" he said in a speech in 1961. "An Oriental poet or philosopher sailing on a primitive sampan boat has no right to laugh at or belittle the material civilization of the men flying over his head in a modern jet airliner." He jokingly called this "syphilisation" – when modernization arrived from the West, you had to accept the bad with the good. In the 1500s, Portuguese traders introduced syphilis into China – at the same time as they brought access to lucrative new export markets for Chinese merchants and producers of tea, silk and porcelain.

Hu was also an accomplished scholar. His books included histories of Chinese philosophy and vernacular literature and Buddhism in China. He directed a project that translated dozens of Western classics into Chinese, including the complete works of Shakespeare. Over the course of his life, he wrote more than 450,000 characters of textual criticisms of Chinese classical novels. He researched *The Dream of the Red Chamber* for more than 30 years. For many, his weakness was that he was interested in too many subjects; so he could not match scholars who specialised in only one. He greatly enjoyed the company of his many friends and this occupied much of his time. Some called him "Doctor Half-Finished" – he completed the first volume of a book, but not the second or the third.

His personal life was just as astonishing. The Chinese moderniser who advocated freedom in love married a woman from a neighbouring village in Anhui; they were betrothed when he was 12 years old. The two had little in common. He made one attempt to divorce her, but her threats and fiery temper dissuaded him; he never tried again. He had long relationships with three other women, one Chinese and two American. If he had been born 20 years later, he would probably have married one of them and not the lady from Anhui.

Most remarkable was his friendship with Edith Clifford Williams, whom he met as a student at Cornell University (1910-1915). Over the next 50 years, they exchanged 300 letters. We have the privilege of reading them because she sent all those she received to the Hu Shih Memorial Hall in Taipei after his death. Hers to him were discovered in the library of Beijing University in 1997 by Professor Chou Chih-ping. With Susan Chan Egan, Professor Chou has written a gripping book on the romance between the two, available in Chinese and English.

With his charm, knowledge and attractive personality, Hu made friends with many people, including U.S. President Franklin Roosevelt and his wife Eleanor, his Secretary of State Cordell Hull and Chinese President Chiang Kai-shek. Lee Ao, one of Taiwan's best known intellectuals, met him and described him in this way– "spring wind and rain that brings life, kept on good terms with everyone." Lee described Hu as a public intellectual who greatly influenced politics and society. If Hu Shih's story can inspire others as it inspired me, then writing this book will have been worthwhile.

CONTENTS

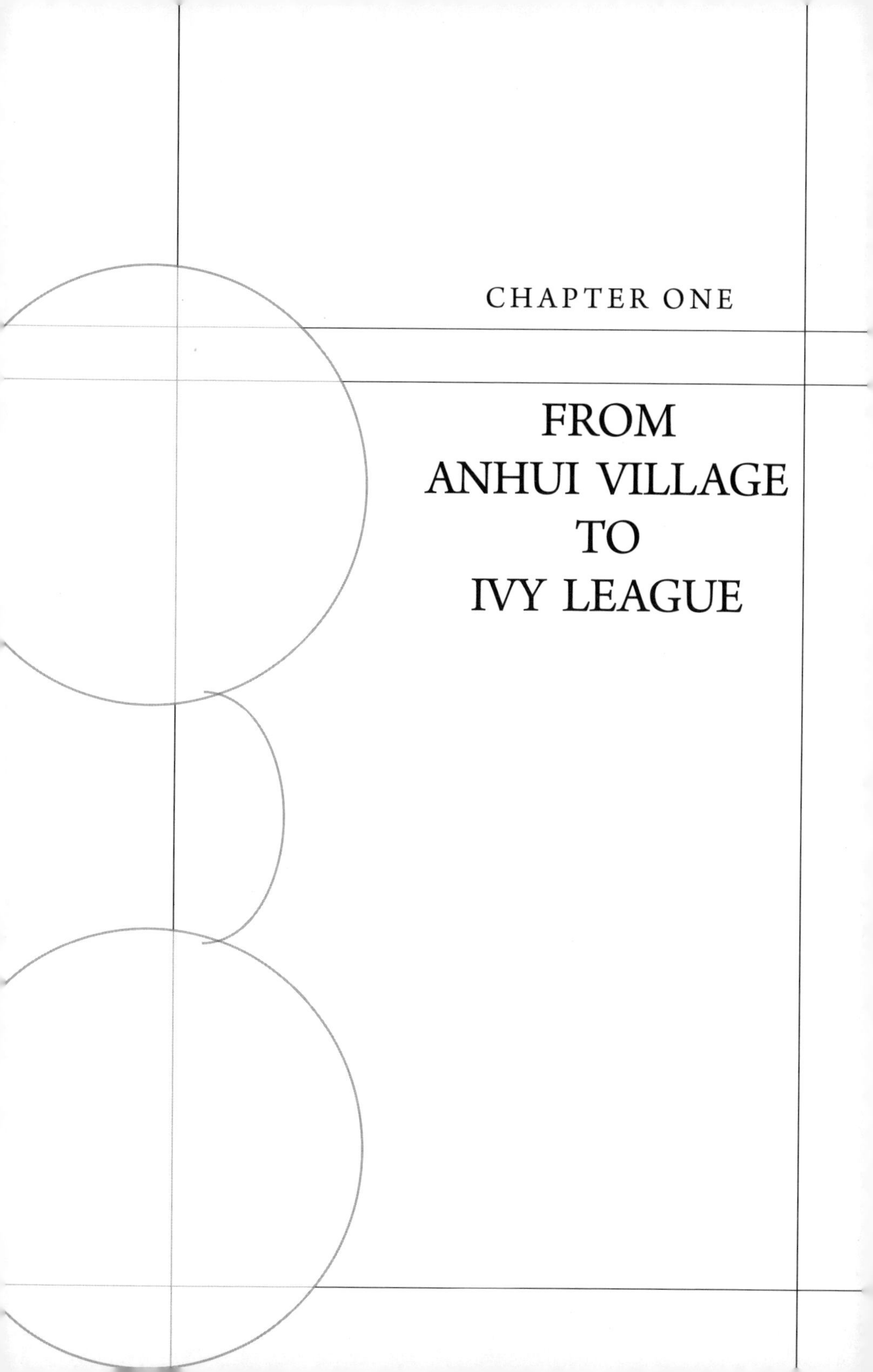

CHAPTER ONE

FROM ANHUI VILLAGE TO IVY LEAGUE

Hu was born on December 17, 1891 in a suburb of Shanghai, into the family of a senior official of the Qing government. His father was Hu Chuan, also known as Hu Tie-hua, superintendent of a district near Shanghai. His mother was Feng Shun-di, his father's third wife. At the time of their marriage in 1889, she was younger than some of the children of her husband's second marriage.

Hu Chuan's first wife was killed in the Taiping rebellion. His second wife bore him six children before her death in 1878. At the age of 48, he married Madame Feng, then 17 years old; his eldest daughter was 28 at that time. Both families came from the county of Jixi, in the southeast corner of Anhui province in East China. Hu Chuan belonged to the elite corps of civil servants who ran the world's most populous country. Like his colleagues, he was highly educated. They served wherever the Qing empire sent them.

In March 1892, Hu Chuan was assigned to Taiwan, as Salt Commissioner of its capital Tainan. His wife and baby son stayed in Shanghai. It was not a posting Hu Chuan welcomed. His health was deteriorating; he had swellings on his legs. Taiwan was remote, backward and famous for its mosquitoes. He would have preferred a job close to Shanghai or his native Anhui.

In April 1893, accompanied by other family members, Madame Feng brought the baby son to Tainan to join him. In June 1893, Hu Chuan was sent to become Magistrate and Garrison Commander of Taitung prefecture in the southeast of Taiwan; he left his wife and son in Tainan. They joined him in Taitung in January 1894.

Despite his busy schedule, Hu Chuan found time to teach Chinese characters to his wife and son. Parents, including those of Madame

Feng, did not consider it necessary to provide a formal education to their daughters. So it was from her husband, and not from a teacher, that she learnt characters. By the time Hu Shih and his mother left Taiwan a year later, she had learnt nearly 1,000 characters and the three-year-old boy more than 700. His father wrote each of them on square red sheets of paper; his mother preserved these sheets her whole life as a precious memento of their short life together.

Their life was turned upside down by the Sino-Japanese war which broke out in Korea in July 1894. Taiwan was placed on a war footing. In January 1895, Hu Chuan sent Madame Feng and their son back to Shanghai and from there, to the safety of the family home in Jixi. China suffered a humiliating defeat. In April 1895, it signed the Treaty of Shimonoseki, under which it ceded Taiwan to Japan.

The Taiwan people were enraged; they established a short-lived Democratic Republic of Taiwan and elected a president. They asked Hu Chuan to command the defence of the southern part of the island against the Japanese army that was soon to arrive. Aged 54, he was in poor health, suffering from beriberi; his legs had swollen and he could barely walk. For this reason, the general of the new 'republic' gave him permission to return to the mainland.

At the end of May 1895, a large Japanese force landed in northern Taiwan. Many Taiwanese resisted the invasion; the Japanese needed a five-month campaign, deploying 100,000 soldiers, before they were able to capture the main cities, including Tainan. Hu Chuan was able to make his way to Tainan. On August 18, he left Taiwan for Xiamen, in Fujian province. He died there four days later.

Later Hu Shih described his father as "the first martyr of the first democratic nation in East Asia". At the time of his father's death, Hu Shih was only three years eight months old and his mother 22. When the family in Jixi read the letter with the news, Madame Feng collapsed with her chair onto the floor; the house was filled with cries of grief. In his will, Hu Chuan told his widow that their son was very intelligent and she must promote his studies.

"Everything I have is thanks to her"

So it fell to Madame Feng to raise their son in Jixi. Born in 1873, she was the daughter of a poor farmer who lived in Jixi not far from the Hu family. In addition to cultivating land, her father was also a tailor. Madame Feng was the first of three daughters; he wanted a son and had to wait for the fourth child for his wish to be realised. As a child, Madame Feng was not taught to read and write. But she realised that education was essential for her young son and devoted herself to this objective. She never remarried and put all her hopes on him.

The love and encouragement of his mother was a key to Hu Shih's success in life. In a letter to an American friend, Edith Clifford Williams, on November 2, 1914, he said: "I had a very good, a very good mother. Everything I have is thanks to her." On another occasion, he wrote: "My mother was very strict with me. She was both doting and strict. But she never once criticised me in front of others and never hit me." She was a devout Buddhist; once she warned him that, if he misbehaved, he could be reborn as a pig or a dog. She woke him at dawn every day and did not allow him to misbehave.

She sent him to a private school in Jixi and paid the teacher more than the other parents, so that he would give him special attention and explain every word and sentence. She guided and corrected him, holding up his father as an example. In his autobiography, Hu later quoted her as saying: "Follow in the steps of your father. He was the only perfect man I ever knew. You should follow after him and not disgrace his memory."

Among the many books he read were those written by his father. By the age of seven, he could read and write with ease. By the age of nine, he had read many Chinese classics. The books Hu read in school were in classical Chinese; it was a foreign language to the students, like Latin for students in Europe. It used words that were 2,000 years old and the teachers had to translate them. To commit them to memory, the students had to recite the texts for hours.

Hu later described this education as "atrocious". While the boys recited the words, the girls were not sent to school; aged six or seven, they had their feet bound. When Hu was nine, he found a copy of a novel written in the vernacular; he read it eagerly. Then he read more novels in the vernacular. Since this was the language spoken in everyday life, these novels were much easier to read than those using classical Chinese. This greatly helped him later in the language reform movement. He wrote later: "Without realising it, I had the practice of reading much vernacular writing. More than 10 years later, this was very useful to me. It helped me write clearly." It was evident to everyone that he had an outstanding intellectual talent. In academic terms, he was soon ahead of the other boys – but small and weaker than they. So his mother told him not to play with them and risk injury.

While mother and son had a close, warm relationship, the same could not be said for the large family among whom they lived. Under the laws of the Qing dynasty, Madame Feng had no rights to the assets of her husband; these went to the male members of the family. She had to depend on them for money to educate her son and for her own expenses.

The other children had opposed the third marriage of their father. The eldest stepson, in legal terms the new head of the family, was a gambler and opium addict who pawned anything he could. Each Chinese New Year creditors came and sat in the house the whole day to demand repayment. So it fell to the second brother to manage the family business. He and the other step-brother ran their shops in Shanghai and Hankou, but neither was profitable.

Hu's mother argued with her two step-daughters in-law, usually over money and the fact that there was less and less of it. A young widow in her 20s, Madame Feng had a low status in the family. She was kind and sweet-natured; if she was angry with her relatives, she kept it to herself. "If I learned a little to be friendly towards others, if I can forgive or excuse others, I have my mother to thank," Hu wrote later. "My clumsy pen cannot describe even one ten-thousandth of the painful bitterness of such a life (of my mother)." In his diary of June 8, 1914, he wrote: "Widowed at twenty-two, she was forced to play stepmother to three grown sons – mere words cannot describe her sufferings."

Madame Feng decided that, to advance academically, her son had to go to a school in Shanghai, China's richest and most developed city. The largest parts of it were the International and French Concessions, which were controlled by the British, American and French governments. They were outside the jurisdiction of the Qing government. This meant that the city had a diversity of schools unavailable anywhere else in China. Some had been established by foreign missionary institutions and others by wealthy Chinese; they were able to offer courses and activities impossible in the state education system.

Madame Feng was only able to take such a decision because the family belonged to the richest class in China, thanks to the salary of her husband.

Hu's three half-brothers had also studied in Shanghai. The family owned a tea shop there, as well as a wine shop in Hankou. The sons of the Hu family were among a tiny fraction of the Chinese population to receive a formal secondary school education.

It was a selfless decision by Madame Feng. She knew that an education in Shanghai would enable her son to fulfill his great potential and acquire skills he could not in Anhui. But it meant that she was losing the centre of her own life, to whom she had devoted all her energy since the death of her husband. It was an exhausting seven-day journey from Jixi to Shanghai. She knew that, once her son had left home, she would see little of him. It was this decision that would allow Hu Shih to join the small elite at the top of Chinese society.

But there was one piece of business she had to do before he left – his engagement. At that time, marriages of young people were arranged by their families. The lady chosen for Hu was Jiang Dongxiu, a native of Jingde county, about 20 kilometres from the Hu family home.

Born on December 19, 1890, Miss Jiang was a year older than Hu. She came from a prominent and well-educated family, which was distantly related to Hu on his mother's side. She was semi-literate and had bound feet. Although Miss Jiang's father had died when she was five, her family was prosperous, while the Hu family was in financial decline. The two families made contact. When Jiang's mother met Hu, she decided he was a very intelligent young boy who would make a good husband for her daughter. The two were introduced. The engagement took place in January 1904; Hu was 12 and his fiancée 13. The marriage ceremony did not take place until nearly 14 years later, in December 1917. A lot would happen before then.

Once that was done, Hu prepared to set out for Shanghai. This is how he described the parting from his mother. "Mother had only me, one person. Her love was too deep and her hope in me too much. Seeing me off to such a distant place, she seemed very excited but would not let slip a single tear." During the next 13 years in Shanghai and the United States, he returned to Jixi only three times. The choice she made was good for him but so bitter for her.

Shanghai – "Before I was 13, I had already become a revolutionary"

The journey to Shanghai took seven days, by foot and boats on inland waterways. Moving there in 1904 was a big shock for the 12-year-old boy from rural Anhui. In his hometown, there was no post office, telegram or newspapers. Shanghai was, after Tokyo, the largest city in East Asia with a population of more than one million people. It was the industrial and commercial capital of China. It had gaslight and electricity; the imposing St Ignatius Catholic Cathedral, with 2,500 seats, was being built. It had a booming textile industry and was an important centre for finance, trade and shipping. It was the centre of journalism and publishing in China – and political fugitives from the Qing government and their revolutionary activity. In the foreign concessions, the Qing police could not arrest them.

On his journey to Shanghai in March 1904, Hu was accompanied by his third half-brother who was suffering from late-stage lung cancer and needed medical treatment in the city. They went to stay in the Rui Xing Tai Teashop, a large establishment run by his second half-brother in the southern district.

Hu discovered that most of the city residents spoke Shanghai dialect, which he did not understand; this was also the medium of instruction in most of its schools. He could speak Mandarin and the Anhui dialect of his native place, which Shanghai people could not understand. He quickly learnt the Shanghai dialect and became fluent in it.

It was an extraordinary period. The city was full of reformist and revolutionary ideas, in part because of its status as a foreign concession. Chinese could say, write and publish things not allowed in the rest of the country. This ferment grew from the widespread belief that this was "la fin de siècle", the final years of the imperial dynasty. Thousands of Chinese had studied and worked in Japan and seen its extraordinary transformation after the Meiji Restoration of 1868. When Hu was a boy, Japan had inflicted a humiliating defeat on China. In 1904-5, it defeated the navy of Tsarist Russia, the first victory of an Asian country over a Western imperial power.

If "little" Japan could transform itself, why not China? In 1820, China had accounted for 32.4 per cent of global GDP, number one in the world. Hu found in Shanghai supporters of Sun Yat-sen, who was leading a revolutionary movement against the dynasty.

Students discussed fervently what kind of country they wanted to build after the end of the Qing – a republic, a constitutional monarchy like Japan and Britain or a different model. The young Hu threw himself into this exciting debate. "Before I was 13, I had already become a revolutionary," he wrote later.

The person who influenced him the most at that time was Liang Chi-chao. Born in 1873, Liang was one of the founders of the Hundred Days Reform movement in Beijing in 1898. He called for the reform of state institutions and the examination system and a constitutional monarchy; he emphasised the importance of individualism. After the conservatives in the government suppressed the movement, Liang had to flee for his life; he moved to Japan where he lived for 14 years. He was a prolific writer; his ideas inspired many young Chinese as well as Hu.

During his six and a half years in Shanghai, Hu studied at three institutions. The first was the Meixi School, whose medium of instruction was Shanghai dialect. Founded in 1878, this was the first school in Shanghai to offer a modern education curriculum, including Chinese literature, mathematics and English. Its founder, Zhang Huan-lun, was a classmate of Hu's father; the two became close friends and, after Hu Chuan's death, Zhang wrote a biography of him.

For the first time, Hu read newspapers; they carried detailed reports of the Russo-Japanese war. He and his fellow students strongly supported their Asian neighbours against Tsarist Russia. Hu already had an excellent foundation in the Chinese language, having read many Classical texts at home. For the first time he read about Japan and its rapid modernisation; if Japan could do this, why not China? He also read revolutionary tracts calling for the overthrow of the Qing dynasy.

While he was there, he lost his third half-brother, who died of lung cancer. Hu stayed at the school for only six months. He and three classmates were selected to take part in civil service examinations organised by the Chinese government of Shanghai. To avoid contact with the Manchu government, the four refused and left Meixi – such was the "revolutionary" zeal of the young men.

Then he moved to Chengzhong Academy. This was another modern school, established in 1900 by a Ningbo businessman. Its first headmaster was Cai Yuan-pei. He became the first Minister of Education in the Republic of China and Chancellor of Beijing University from 1916-27; he first met Hu at the school and hired him to work at Beijing University after he returned from the U.S. He was, like Hu, a pioneer in the reform and modernisation of China.

At Chengzhong, Hu concentrated on English and mathematics and made rapid progress in both subjects. He thrived under its strict management and personalised tuition. He studied there for 18 months. One of the books he read there was "Evolution and Ethics" by Thomas Henry Huxley, which had been translated into Chinese in 1898; it was widely read by middle school students in China. Huxley was an English biologist and anthropologist who argued the Darwinian case for the survival of the fittest and applying a scientific scepticism to all subjects.

Hu and his classmates took from it the message that, to survive in the international world, China would have to adapt and change radically. It inspired him to change his name. His second step-brother proposed Shih, meaning "fit, suitable, proper". In his application for the exam to study in the United States, Hu used that character – and kept that name for the rest of his life.

In the summer of 1906, he moved to the school that had the biggest impact on him of the three he attended in Shanghai. This was China National Institute (CNI), which had a remarkable history. At that time, Japan was the favoured destination for Chinese students going abroad. In November 1905, the Japanese Ministry of Education issued new regulations that severely restricted foreign students; Koreans and Chinese were the majority. The 8,000 Chinese students in Tokyo went on strike to protest; one of them wrote a 5,000-word suicide letter and jumped to his death in the sea. Then 3,000 of the Chinese left their colleges and returned home.

After arrival in Shanghai on December 21, they determined to set up their own college; they called it China National Institute. It opened its doors on April 10, 1906 on Sichuan Bei Lu, north of Huang Ban Qiao, with 318 students.

For Hu, it was a great good fortune to study there for more than two years. Most schools in the city used Shanghai dialect as the teaching medium, but CNI used Mandarin since it had students from all over China. There Hu met students from many parts of the country, many older than himself; there were just 12 from his native Anhui. These encounters gave him an understanding of conditions throughout China and the situation in Japan. Some classmates become good friends.

He studied poetry and vernacular Chinese; he became editor of a student magazine that published articles written in the vernacular. This was excellent practice for the language revolution he would launch after he returned from the United States. The articles he wrote covered themes he would pursue in later life – such as an end to foot-binding and lack of education for women. He was active in student affairs and meetings. From 1928 to 1930, he would be president of the school.

From the end of June to early September in 1906, he returned to the family home in Jixi. He was suffering from beriberi, an illness caused by a deficiency of thiamine, vitamin B1. He spent the time at home recovering from his illness. His mother was delighted to see him and academic progress he had made in many subjects. Hu kept himself busy by reading and writing poetry; he realised that he had a deep interest in literature.

Night in a police cell, missing a shoe

But the honeymoon at CNI did not last long. From the end of 1907, the students and the directors began to argue over its management. In the autumn of 1908, unhappy that they had lost control, some students left CNI and set up another establishment, New China National Institute.

Hu was among them – but was unable to study there because the family money had run out. His father's death had left the family several thousand taels of silver, a substantial sum. But all had been spent on Hu's studies and subsidising the two unprofitable businesses in Hankou and Shanghai.

The second half-brother who managed the Shanghai tea shop was a spendthrift who neglected the business. Finally, he ran out of money and was forced to hand it over to his creditors. So Hu could no longer stay there or at the school. He went to live in the office of the magazine he was editing; he ate in its canteen. He supplemented his modest salary by teaching two classes of English at a Shanghai secondary school. He read the novels of Charles Dickens, Victor Hugo and Leo Tolstoy in English.

In November 1909, he stopped teaching and moved with friends into an apartment in Haining Road. In early 1910, Wang Yun-wu, his former professor at CNI, introduced him to a new school, Hua Tong Public School, as a teacher of Chinese. It was established to educate the children of the poor. Hu found the work there difficult. The students had received no education at home; they were rough and difficult to manage. He felt depressed and frustrated, falling into a lifestyle uncharacteristic of the disciplined young man who had come such a long way from rural Anhui. With his friends, he drank heavily, played mah-jong and did not sleep regularly. On the evening of March 16, they spent the night at a brothel – and he staggered home at 0600 to correct his students' homework and go to class.

He was suddenly awakened from this lifestyle by an incident soon afterwards when he attacked a policeman and spent a night in a police cell. He woke up the next morning with a dense hangover, dirty and dishevelled and missing a shoe. He asked the officer what had happened. The officer explained that he was on patrol at midnight on Haining Road and found Hu shouting and waving a leather shoe. When he shone his torch, Hu began to abuse him, calling him a "foreign slave" – a reference to the fact that he was part of a British-led police force. Hu then hit him with his shoe; the two started to fight and fell to the ground. It was pouring with rain and the ground was very slippery. The officer summoned a passing horse-cart; the two drivers helped him to overpower the drunk Hu and take him to the station.

He was a lucky man. When asked his name and profession, Hu said that he was a teacher at the Hua Tong Public School, which was well-known in the district. So the officer only issued a fine of five yuan, to cover his injury and the torch which Hu had broken, and took no further action.

Hu felt great remorse over the incident; he had let down his mother and vowed not to repeat such a mistake. He drastically reduced his drinking and turned his attention to a new project. Since 1908, he had the dream of studying in the United States, which had suddenly become possible because of a rare example of enlightenment by a foreign power.

In 1906, Edmund J. James, president of the University of Illinois, proposed to President Theodore Roosevelt a plan to establish a scholarship programme to send Chinese students to the U.S., using money the government was receiving from the Boxer Indemnity. The U.S. share was 7.32 per cent of the 450 million taels of fine silver China agreed to pay from 1901 until the end of 1940; it was one of 11 foreign nations to receive payment, as compensation for the human and property damage caused by the Boxer Rebellion of 1900.

Hu Shih in 1909

"China is upon the verge of a revolution," James wrote in a letter to Roosevelt. "The nation which succeeds in educating the young Chinese of the present generation will be the nation which for a given expenditure of effort will reap the largest possible returns in moral, intellectual and commercial influence."

Roosevelt accepted the proposal. At that time, the other western powers were busy maximising their commercial and diplomatic privileges in China; none thought of a similar gesture. The U.S. launched the programme in 1909; the money funded the selection, training, transport and study of those who qualified. That opened the door to young men like Hu whose families did not have the substantial funds needed to send their children to study in the U.S. 1910 was the second year of the programme, with the exam to be held at the end of July. Hu resigned from his teaching job and devoted the next months to prepare for the exam.

Between 1909 and 1929, the scheme sent around 1,300 Chinese students to the U.S. The exam was highly competitive, with less than 10 per cent of applicants qualifying in some years. The programme produced many distinguished Chinese, including Nobel Physics prize winner Yang Zhen-ning. Hu was one of the most outstanding results, who made major contributions to both China and the United States.

Hu was greatly helped by a close friend named Xu Yi-sun who encouraged Hu to sit for the exam and provided the money he needed. This included regular payments to his mother, repayment of a small debt and the living costs for two months of study and travelling to Beijing to sit for the exam.

Also from Jixi in Anhui, Xu studied with Hu in the same class as CNI; the two men lived together. Xu himself went to study in Japan in 1913, returning home in 1916. Sadly, just three years later, he died. Hu wrote a long emotional article in memory of him; he praised his sincerity and loyalty, saying it would be hard to find a better friend.

The topics covered in the Boxer Indemnity Scholarship exam would have taxed a student of any nation: Chinese and English; ancient and modern history of the West including the history of Greece and Rome; Solon and Lycurgus; one foreign language, French or German; Latin optional; algebra and geometry; physics, chemistry and botany; and animals.

It was based on the exam given to those in the United States applying to enter universities. On the evening of June 28, Hu and his second half-brother took the railway – a novelty in China – and arrived in Beijing on July 3. Thanks to a friend of his half-brother, Hu was able to stay at a school. The first part of the exam was Chinese and English. Hu excelled at both and finished 10th out of the 150 sitting the exam. He was not so strong in the second part, which included the other subjects. He had not studied Latin, French or German. In the second part, he finished in 55th place. But, since there were 70 places, he had done well enough and was selected to go. Unfortunately, there was no time to return to Jixi and say goodbye to his beloved mother. On August 16 1910, Hu and 68 others – one had dropped out – left Shanghai on the "China" ship for San Francisco. The vessel belonged to the Pacific Mail Steamship Company of the United States.

It was the most important moment in Hu's life. His seven years in the United States enabled him to have a life he and his mother could never have imagined. If he had stayed in China, he would probably have become a school teacher, a publisher or civil servant. His success in the exam was due to his intellectual ability, years of determined study and the faith and sacrifice of his mother.

The six and a half years at three different schools in Shanghai were an unusual experience for a young Chinese. For centuries, education – such as that received by Hu's father – had consisted of rote learning and strict obedience to parents and teachers; the teachers spoke, the pupils listened and took notes. The curriculum had been unchanged for decades. But Hu and his classmates took subjects that had never been on the traditional curriculum. They felt emboldened to change school twice.

The CNI was the most unusual of the three, set up by students who

In August 1910, a group photo in Shanghai of the Chinese going to study in the United States. Hu Shih is on the far left of the second row.

wanted control of the school. They held fierce debates and were willing to defy their superiors. This was excellent preparation for the world Hu would encounter on the other side of the Pacific. Such an experience was only possible because of the special status of Shanghai, a city in China but controlled by British, Americans and French. They were willing to give freedom and space to these new modern schools and allow journalism and publishing to flourish.

Hu and his classmates knew that the Qing dynasty did not have long left. The humiliations of the defeat by Japan and the Boxer Indemnity were signs that it was not fit to govern. The young men believed they must make best use of their time and prepare for the new country that would replace the Qing dynasty.

Sources for Chapter One

Young Hu Shih 1891-1917, by Tang Yan, Spring Hill Publishing Company, Taipei city, first edition June 2020.

If Not Me, Then Who? Hu Shih Volume One, Jiang Yong-zhen, Linking Publishing Company, Taiwan, first edition January 2011.

Oral autobiography of Hu Shih, as told to Tang De-gang, Yuanli Publishing Company of Taiwan.

Hu Shih and Female Emancipation in China, by Harriet Chien-ming Twanmoh, published by Australian National University, 1966.

A Pragmatist and his Free Spirit, the Half-Century Romance of Hu Shih and Edith Clifford Williams, by Susan Chan Egan and Chou Chih-p'ing, Chinese University Press of Hong Kong, 2009.

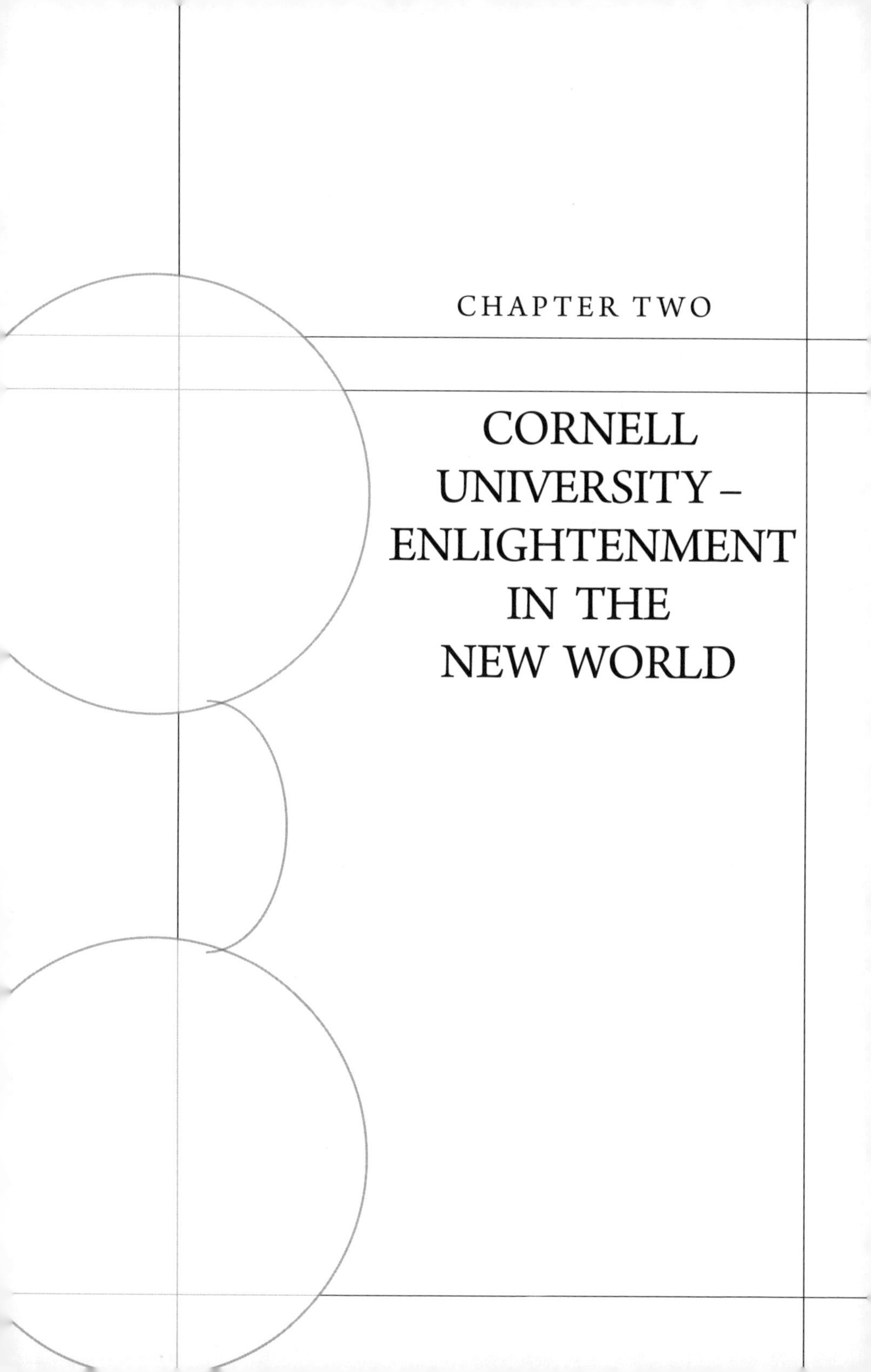

CHAPTER TWO

CORNELL UNIVERSITY – ENLIGHTENMENT IN THE NEW WORLD

The vessel carrying Hu and his fellow students to the U.S. stopped at Nagasaki, Kobe and Yokohama; it gave them a brief opportunity to see these Japanese cities. Those travelling with Hu described him as thin and not so healthy looking, but full of self-confidence – and a lover of debates and playing cards.

One issue each student had to decide was what to do with his pigtail; it was mandatory for men in the Qing empire. But they knew that, in the United States, it would make them subject to ridicule. Before he left, Hu has his pigtail cut off; he sent it to his mother for safe-keeping.

Before their departure, many groups in Shanghai gave the students farewell banquets; people regarded them as ambassadors for the nation. This warm reception continued in Hawaii, San Francisco and the cities through which they passed on their rail journey to the East. Associations of Chinese students feted them – the best way to be welcomed in this alien country.

How blessed they were by comparison to their compatriots who went to the U.S. as workers. The Chinese Exclusion Act of May 1882 had banned immigration of Chinese labourers. The law was a result of widespread racism and resentment of white Americans against Chinese willing to work for lower wages and to endure worse conditions.

Earlier legislation in 1875 had banned Chinese women from emigrating to the U.S. But the law allowed entry of Chinese diplomats, teachers, merchants and students, such as Hu. The exclusion act was not repealed until 1943.

Hu and his fellow students were blessed to live and study in university towns whose residents were more open-mind; they did not experience the racism endured by their compatriots in Chinatowns in San Francisco, New York and other large cities.

On September 18, 1910 Hu reached Ithaca in upstate New York, home of Cornell University. It was accepting 19 Chinese students, the most of any American university from the Boxer Indemnity group that year. Hu was fortunate to enter one of the most open-minded institutions in the U.S. A private university established in 1865, it was one of the few to admit students of different colours and nationalities. By 1870, it had students from 28 states and 11 foreign countries, including Japan. By 1908, there were 53 Japanese alumni, 37 Chinese, 26 Filipinos, 16 Indians and one Korean. In 1866, it admitted its first African-American students.

Another singularity was that it accepted lady students. As a result, Hu was able to meet a wide diversity of men and women, an experience impossible for nearly all Chinese – and, indeed, the large majority of Americans.

The year he arrived, Cornell had 5,000 students, including 700-800 women. The university had two dormitories for the ladies but not for the men. So Hu and seven other Chinese rented rooms in the house of an elderly woman in the small town. In an article he wrote in 1915, he said that, when he left China, he was a nationalist from head to toe. "Then I met delightful South African, South American, Filipino, Japanese and Jewish people and had close contact with them. I was able gradually to throw away the prejudices of my early years," he wrote. Between 1900 and 1949, 3,500 Chinese students enrolled at Cornell.

Hu and the other students were also blessed by the welcome they received from the university faculty and their families; they considered it their duty not only to teach the newcomers in the classroom but also to welcome them into their homes, churches and social lives. To understand the U.S. meant both learning in the lecture hall, library and laboratory and joining American society. This was, and is, not an opportunity given to all students who go to foreign countries. It was one that Hu seized with both hands. The warmth of this reception and the excellent access he was given were major reasons why he regarded the United States with such favour throughout his life.

Study Something Useful

Hu's family wanted him to study railway or mining engineering – knowledge that would enable him to earn a good living and help China. They urged him not to study literature, philosophy or law, which they considered not useful nor profitable for the future. But he was not interested in the subjects picked by his family; instead, he chose agriculture. "I felt that, with agriculture, I could contribute to China," he said in Taiwan late in his life. Cornell was famous for its agricultural faculty, with 1,230 students during his first year, the largest number of any university in the U.S. It had the additional benefit of being free – so Hu could save some of his scholarship money and send it to his impoverished mother. He had brought in his luggage 1,300 silk-bound books; he intended to continue reading the material he had enjoyed in China.

In his first academic year, his curriculum included botany, biology, meteorology and chemistry – subjects necessary for agriculture. To these, Hu added English, which included the plays of Shakespeare, such as *Henry IV* and *Romeo and Juliet*. He also took courses in German, enabling him to start reading literature in that language.

In February 1911, on his own, he started to learn Latin and, in the summer, Greek – neither were in his course schedule. These choices were

Hu Shih at Cornell University

an indication of his intense intellectual curiosity and discipline, to find time for these in a crowded academic and social schedule. He also read the books he had brought from home. In the first-year exams in June 1911, he scored 89 points in the English One exam, 82 in Biology One and 80 in Plant Study and German Two.

On September 28, the autumn term began. Just two weeks later came an event that shook the world – the Xinhai revolution of October 10 in Wuhan and the end of the Qing dynasty after 268 years. The Chinese students were overjoyed at the news, if apprehensive over what would come next. The event aroused great interest among Americans; they were eager to know its significance and its impact on the U.S.

Hu was among the students invited to talk to American audiences about the revolution; he was eager to do this and practise public speaking in English. It was an excellent opportunity to learn how to interact with Westerners, a skill that he would use throughout his life. So, in 1912, he enrolled in classes in public speaking; he gave addresses in churches, learned societies and women's groups. He regarded it as a good reason to research current affairs, improve his writing skills and present his ideas in a clear and concise way.

The Qing government was replaced by the Republic of China. Its first president, Sun Yat-sen, was inspired by many of the same ideas that Hu heard in Cornell. Sun received five years of education in Hawaii andeight in Hong Kong. His political creed was the "Three Principles of the People", which owed much to American ideals. He said his philosophy was inspired by the Gettysburg Address of President Abraham Lincoln in November 1863: "Government of the people, by the people, for the people".

He advocated a separation of powers between executive, legislature and the judiciary, as in the United States. But Sun was president for less than three months, stepping down on March 10, 1912 in favour of warlord Yuan Shi-kai. The new republic became a battleground of rival warlords; the calls for democracy and constitutional government were drowned out by the sound of rifles and artillery.

No More Apples, Please

Back in Cornell, Hu had realised that he should not be studying agriculture; it was not his interest. Unlike the vast majority of his countrymen, he had not grown up on a farm; instead, he had been surrounded by family heirlooms, books collected by his father and IOUs signed by his half-brother. Then he spent six-and-a-half years in Shanghai, the most urban place in China. In addition, what he was learning was the agriculture of the United States – how to raise cattle, sheep and horses, farm large plains of corn and wheat and grow apple and citrus trees. This was not the agriculture of his neighbours in Anhui.

In his memoirs, he described this disconnect most vividly in trips to study 30-35 different types of apple; his assigned subject was pomology, the study of apples. Brought up among apple trees, American students could easily spot the difference between the varieties, even without cutting them. The Chinese students had to dissect the apples in the laboratory; even then they could not name all the varieties.

"As a young man, I had a good memory and could take the night train before an exam and remember all the varieties for an exam. But I knew that, three to seven days after the exam, I would forget all the 400 varieties (of apples) completely. Half of them do not exist in China. So, I decided that studying agriculture was against my interests and character, basically stupid and a waste of time," Hu wrote.

Since his childhood, he had enjoyed literature; he had enriched this experience by studying English, French and German literature at Cornell. Another reason to switch from agriculture was the Xinhai revolution; the talks he was giving required him to research modern history and current affairs.

So, in February 1912, he moved to the faculty of philosophy. The subjects in his new schedule included: fine arts; ethics; the Renaissance; principles of business management; political institutions and comparative politics; French; English drama up to 1642; and Victorian poetry.

He was much more at home. In the summer term, he added public speaking: public finance; and history (including the growth of the British Empire and the Napoleonic era). During the next two years, his courses included: history of ethics; reading of German philosophy; critical philosophy of Kant; and empiricism and rationalism.

At Cornell, he obtained a M.A. and a B.A, in 1914. Then he began graduate studies in philosophy. He liked Ithaca, home of Cornell; it was a medium-size town in the Finger Lakes region of New York state, 360 kilometres northwest of New York City. He enjoyed the scenery, with hills and lakes, and the tranquility. t night, people did not lock the doors of their houses. But, a native of south China, he found the cold and winds of winter hard to bear.

Cosmopolitan Club and "New Pacificism"

The Cornell website describes Hu's many activities outside the classroom. In the 1910-11 year, he joined the Chinese Students' Club and the Cosmopolitan Club(CC). Founded in 1904, CC was the first group in Cornell for international students; it constructed its own building at 301 Bryant Avenue in 1910. It hosted lectures by faculty and students and organised social events and exchanges with other universities.

Hu remained active in the CC during his time at Cornell, serving as president in the 1913-1914 year. He regarded it as a precious way to meet students from all over the world and enrich his knowledge of their

During his time at Cornell, Hu Shih (second left) goes on an outing with his fellow students.

countries. "This was one of the most important harvests of my studies in the United States," he said. He took part in national meetings of the CC in Cornell and other universities. For three years from the summer of 1911, Hu lived in the CC building. In December 1913, he was a delegate to an International Congress in Washington; its members were received by President Woodrow Wilson and Secretary of State William Jennings Bryan – his first experience of U.S. politics.

In the 1912-13 year, he joined the fraternity Phi Beta Kappa and helped to organise the Ninth Annual Conference of the eastern section of the Chinese Students' Alliance of America in Ithaca.

In May 1914, he received a Graduate Scholars in Philosophy award and the Corson Browning Prize, a gold medal worth US$50 from the English department. This was for an essay on Robert Browning; he was the first Asian to win the prize – the news merited a story in the *New York Herald* and many other American newspapers. The money was useful to help cover his living expenses and regular payments to his mother.

In the 1913-14 year, he was home news editor of the *Chinese Students' Monthly* magazine. His wide knowledge and fluency in English made him a popular speaker for public events – about 70 in the three years to March 1915. The topics included "Confucianism and Taoism", "Chinese Women", "The Christian Opportunity in China" and "China's Entrance into the (First World) War." All were in English. Hu immersed himself in U.S. social and political affairs and came to know them better than many Americans.

Hu and the other Chinese students donated to Cornell 300 Chinese-language books, on literature, the economy, Confucianism, Taoism and history, by famous authors. The books had an English translation and summary. They became part of what became the university's large Wason Collection. "It is our great pleasure to see the library of this university become one of the greatest college libraries in America," he wrote. "It is also our duty to do our best to help the library grow." This collection was named after Charles Wason who made the greatest single donation in 1914. Today it comprises more than 608,000 monographs, including 373,000 in Chinese, 147,000 in Japanese, 10,500 in Korean and 78,000 in Western languages.

World War One & "The Great Illusion"

During his meetings at the CC, Hu encountered the most famous pacifist of his generation. This was Norman Angell, a British journalist and author of *The Great Illusion,* published in 1909. It was translated into many languages and had a global impact. Many people were looking for an alternative to the conflicts between imperial powers that dominated global affairs: was there no other way to manage global affairs?

In 1933, Angell received the Nobel Peace Prize in recognition of this book. His thesis was that the integration of the economies of Europe had reached the point that war between them would be futile; this had made militarism obsolete. The outbreak of World War One in Europe proved him wrong – but the devastating loss of life gave additional energy to his movement, which came to be known as "New Pacifism".

In 1915, his supporters in the U.S. set up the International Polity Club (IPC). It held two international meetings, at Cornell University in 1915 and in Illinois in 1916. Hu attended both; at the first, he met Angell and heard him speak. One result of these meetings was Hu's response to the world war in China.

Wisely, the Chinese government had declared its neutrality in the war; it did not want the fighting to extend to the 40 concessions held by foreign countries within its territory.

The most aggressive of these powers was Japan, with eight concessions. In January 1915, Japan presented 21 demands to the young Chinese government; these demands would greatly extend Japan's control over Manchuria and the Chinese economy. The demands provoked a public uproar and a nationwide boycott of Japanese goods; its exports to China fell by 40 per cent.

This indignation was shared by Hu and the other Chinese students at Cornell. Most called for war with Japan. Hu was among a tiny minority calling for calm and reflection. He knew well that China would certainly lose a war with Japan, as it had done 20 years earlier; this had hastened the death of his father. In an open letter to the *Chinese Students' Monthly*, he wrote: "Let us be calm, let us do our duty, which is to study ... the final solution of the Far Eastern Questions is not to be sought in fighting Japan at present, not in any external interference by any other power or powers."

His letter provoked strong, sometimes violent, disapproval. In letters to newspapers, he further explained his views: "In this 20th century, no nation can ever hope peacefully to rule over, or interfere with, the internal administrative affairs of another nation. A Japanese attempt to assume control of China will result in a sea of trouble."

How accurate and foresightful he was. In his diary for February 21 1915, he wrote: "It is no disgrace for a nation to lack a navy or an army. It is only a disgrace for a nation to lack public libraries, museums and art galleries. This is the disgrace of which our people must rid themselves."

Despite their disagreement with his views, his fellow students remained friends with Hu and chose him as editor of magazines by Chinese students in the U.S. In 1920, after the League of Nations was founded in Geneva, Hu was a founder of the Chinese Association of the League of Nations.

He included among his friends Japanese students. Rare among Chinese, he sought their company in order to understand their point of view. He knew that China had much to learn from its neighbour in how to modernise and catch up with the West. It was one of the great tragedies of the 20th century that the leaders of China and Japan were unlike Hu; they were unable or unwilling to listen to each other and resolve their differences peacefully. The wars they fought led to catastrophe in both countries.

In 1916, the American Association for International Conciliation gave Hu an award and US$100 in prize money for his essay "Is there a Substitute for Force in International Relations?" It was translated into Spanish and Portuguese. The essay advocated a system whereby different countries co-ordinated their military forces to enforce peace and international law.

Edith Clifford Williams, Lifelong Friend

Hu had another important experience at Cornell – meeting Edith Clifford Williams, second daughter of Henry Shaler Williams, professor of geology at the university. She was a student in the arts faculty, a year behind Hu. The two became lifelong friends; over the next 50 years, they exchanged over 300 letters, along with poems and excerpts from Hu's diaries and other documents. The two met in the summer of 1914, when Hu was in his first year of graduate studies in philosophy at the Sage School. He was 23; she was 29 and unmarried. In the U.S. of that time, most women married in their late teens or early 20s; the common view was that "she had missed the boat".

Her name first appears in Hu's diary of October 20 that year; he described a three-hour country walk with her: "A very intelligent and widely read person, eccentric. She pays no attention whatsoever to clothes and jewels even though she comes from a wealthy family. One day recently she decided to cut her own hair so short that only two to three inches were left. Her mother and older sister were appalled."

Edith Clifford Williams aged 16

What impressed Hu were Williams' intellect, education, individuality and unorthodox opinions. She was unlike any woman he had met before; she was intelligent, well-read and regarded herself as the equal of men. During his six years in Shanghai, Hu had little contact with women of his own age. In his early years at Cornell, most of the American women he met were middle aged or elderly, usually the wives of professors and those involved in churches and civic societies.

In his diary for June 8, 1914, he wrote: "It is not too late for me to attend to my emotional development. I must take advantage of being in this country and being at a co-educational university to get to know some educated women." He wanted to make up for lost time. It was his good fortune that Cornell was one of the few colleges in the U.S. with female students. His self-confidence, sociability and fluency in English enabled him to mix with them.

Williams was born on April 17, 1885. She came from a wealthy and well-connected family. From the age of seven to 19, she grew up in New Haven, where her father was a professor at Yale University; in 1904, he transferred to Cornell and brought his family with him.

His views were avant-garde. He told Edith to avoid marriage if she could. He doted on her and encouraged her to live as she wished. In 1903-04, she enrolled at the Yale School of the Fine Arts and studied with Impressionist painter John Henry Twatchman.

Despite the opposition of her mother, she went on her own to study art in New York; most of her friends were poor artists. In May 1906, she went to London to study art. Her mother opposed her being on her own there without a chaperone. In November, she enrolled briefly at the Academie Julian in Paris, before returning home early the next year. She joined the avant-garde art movement, in oil painting and sculpture. In 1917, she showed two works at the inaugural exhibition of the Society of Independent Artists in New York. One of them, *Two Rhythms,* is on permanent exhibition at the Philadelphia Museum of Art.

Hu probably first met her at the house of her parents. The friendship developed rapidly after they attended a wedding together in Ithaca in June 1914. Hu became a regular visitor to the Williams' home in Ithaca, including for Thanksgiving Day dinner: Edith's mother liked him.

Hu and Edith became regular correspondents, on many subjects. In his diary, Hu wrote: "In Miss Williams's place, I saw a huge bundle of letters that I had written to her over the past two and a half years ... Since knowing my friend Miss Williams, my ideas toward women have greatly altered. I believe in female education. To bring up a good wife and mother for the country is a preparation for family education. Now I realise that the highest aim is that of bringing up a new kind of independent woman. A nation which possesses 'independent women' will be able to improve its people's morals and personalities."

Miss Williams was one of three highly educated women Hu befriended at Cornell. Each impressed him by their education, independent thinking and ability to make a life outside the traditional family. Miss Williams's opinions greatly influenced Hu's early social and political ideas.

In a letter dated January 30, 1915, from her apartment in New York, this is what Miss Williams wrote to Hu about sex: "If the truth of sex attraction is clearly understood and valued for just so much as it is good for, and, when it consciously appears not of use, it is consciously put away by willful turning of the attention to the higher side of that friendship."

This was a time when public discussion of sex was taboo in Western as much as in Oriental society; and here was a foreign woman revealing to Hu her thoughts on the subject. The two had to behave with care; Williams' mother strongly objected to her daughter and Hu being alone in her New York apartment. She was a conventional woman, at a time of strong prejudice against Asians among Americans – but Miss Williams never regarded Hu as a 'Chinaman'.

Hu wrote regularly to his mother and hid nothing from her. In a letter to her dated February 18, 1915, he wrote that, of all the women he had met in the U.S., Williams was the closest to him. "She is highly intelligent, kind

Edith Clifford Williams with her family: she is on the far right.

and knowledgeable and I benefit very much from her views. I have told her about you and she is full of admiration … I was very happy to receive the picture of Dong-xiu (his fiancée) you enclosed with your last letter."

In return, his mother sent a letter to Williams, which Hu translated for her. "You have helped cutting and chiseling my son's thinking," said Madame Feng. "As I turn my head toward the western heavens, I long very much to know you. But I can only send this note to thank you with all truthfulness and earnestness and to wish you infinite happiness."

With this honesty, Hu achieved two objectives. One was to reassure his mother that he remained committed to the lady she had chosen as his wife. The other was to tell Miss Williams, and other Americans, that he was "not available".

"I almost became a Christian"

For the Christian churches of the United States, foreign students offered a fertile ground for evangelism. These well-educated young people were away from their families and communities and the rules that bound them; they were ready to hear a new voice. For the churches, Hu and his compatriots were no exception – the evangelisation of China, the world's largest "heathen" country, was one of the great missionary projects of the 20th century. The churches actively sought out the new arrivals; they offered them material and spiritual help to adapt to this new and alien land.

In the summer of 1911, Hu attended a summer camp of the Chinese Christian Students Association of North America in Pocono Pines, Pennsylvania. For a year and a half, he attended Bible studies with William W. Comfort, a Quaker and Professor of French. He read the Old and New Testaments and greatly admired the content; he also met Jews and Mormons. Many around him, Americans and Chinese, were devout Christians. "In my diary and letters to my friends, I said that I almost became a Christian. But, afterwards, I went back on my word," he said in his oral biography in later life. Hu was moved and interested by what he read and heard but did not convert. He was, and remained, someone who believed in science and what could be proved; he did not believe in the supernatural.

Like any foreigner, Hu was impressed by the power and influence of

Christianity in the United States. The country had been founded by intensely religious people who wanted the freedom to practise their faith away from the constraints of King and Church in Europe. The "Protestant work ethic" seemed to be one reason for the country's rapid industrial and scientific progress. How had it pulled so far ahead of the countries of Latin America that were also new and founded by migrants from Europe but were Roman Catholic? What could China learn from its example?

A New Language for China

It was during the summer of 1915 that Hu and his fellow students began to debate an issue that would profoundly influence their native country. Should classical Chinese be retained as the written language or should it be replaced by the vernacular spoken by ordinary people? Should the characters be written in Roman letters that were easier to learn?

It was a continuation of the debate Hu and his classmates had begun at secondary school in Shanghai. He had written articles in the vernacular for the student magazine at China National Institute. This debate took on a greater importance after the end of the Qing dynasty and the foundation of the Republic; this made possible reforms that had been impossible 10 or 20 years earlier.

Hu and his classmates belonged to the small elite who had an opportunity for education and travel that was impossible for the vast majority of their fellow Chinese. They knew that, when they went home, they must give back something from the learning they had received.

The revolution of 1911 provided them with a rare opportunity to change their country. This was not given to Chinese who had returned home after studying overseas during the Qing dynasty. At that time, many of those who had proposed reform had been imprisoned, killed or forced into exile. But now the Republic had been established; many things had become possible. The old regime was dead and the new one was only starting to take shape.

Classical Chinese was the written form of the language that had been used from the end of the Han period (AD 220) through the dynasties that followed, including the Qing. It was also used, during various periods, in Japan, Korea and Vietnam. The Chinese government used it in official documents. To pass the civil service exam, a man had to master the language which he would write during his time in public office.

But, while the government continued to use the same written form, the spoken language used by ordinary people evolved over time; the gap between the two forms became wider and wider. Only a limited number of officials and the educated class – a small percentage of the population – could understand the written language.

The issue was further complicated by the fact that people in different regions of China used dialects that were mutually unintelligible. While the characters were the same all over the nation, the way they were pronounced varied dramatically between Guangdong and Shanghai, Hunan and Sichuan.

When Hu went to study in Shanghai, he found that the teachers used Shanghai dialect as the medium of instruction; he had to master it in order to follow the lessons.

During the last two decades of the 19th century, reformers began to question the use of classical Chinese and the enormous gap between the intellectual elite and the common people. If the large majority of people could not read written material and were illiterate, how could China become a modern nation? Some argued that, because of its complexity and the time needed to learn the characters, the Chinese language itself was an obstacle to modernisation and new learning in science, medicine, engineering, industry, the military and other fields.

A similar debate had started more than 20 years earlier in Japan, after the Meiji Restoration of 1868. The situation was like that in China. Classical Japanese began to be written during the Heian period (794-1185 AD), when it was very close to spoken Japanese. It became the standard for writing; the spoken language evolved and changed, but the written one did not. So, by the 19th century, the two forms were greatly different. As a result, a majority of people could not read the classical Japanese used in official life.

Reformers put forward many ideas – simplify the language, use fewer Chinese characters or even abandon them completely and translate them into a Roman alphabet. There were even proposals to replace Japanese with English or French – the languages of the world's two most advanced nations.

In 1903, the National Language Research Council was set up; it proposed that the Tokyo dialect be adopted as the national standard. This was a written form of the colloquial language spoken by upper-class residents of Tokyo, the capital; this became the new national standard. From 1903, the Ministry of Education began to produce textbooks in this style; they started to be used in primary school textbooks two years later. The Japanese expression for this was "gembun itchi)",which means "unification of spoken and written". It included Chinese characters and words written in the two Japanese alphabets, hiragana and katakana. This has become the standard Japanese used until the present.

By 1908, novels were written in this style, not classical Japanese, and, by the 1920s, all newspapers. But government documents were written in classical Japanese until 1946, after the defeat in World War Two. In this, as in many fields of modernisation, Japan set a good example for its giant neighbour.

During his last four years in Shanghai, Hu had enthusiastically entered the language debate. He became editor of the student magazine at China National Institute and published articles in vernacular Chinese.

Hu was a reformer, arguing that this vernacular form should become the standard for written Chinese; this was the democratic choice, because it would give millions more people access to learning and knowledge. Not all his fellow students agreed. For them, classical Chinese expressed the beauty and historical knowledge of the nation and should not be abandoned. There were fierce debates between the two sides. Hu developed his ideas on this subject at Columbia University.

The nearest equivalent in Europe was the use of Latin as the official language of the Roman Catholic church. This began in the late fourth century AD. Many languages of western Europe are based on Latin. But they changed and developed; the language used in the church remained the same. Ordinary people could not understand Latin unless they had studied it at school or seminary.

One of the main tenets of the Protestant Reformation from the 16th century was the use of local languages, not Latin, in church services, so that everyone in the congregation could understand them. It was only at the Second Vatican Council of 1962-65 that the Catholic church approved the use of local languages.

From Cornell to Columbia

In 1914, Hu entered the graduate programme in philosophy at Cornell. During this time, he discovered the writings of Professor John Dewey of Columbia University on experimentalism. In the summer of 1915, he did a systematic study and research of Dewey's work and decided to transfer to Columbia University in New York to study under him.

The desire to study under Dewey was one reason for the move. Another was that the Cornell department had withdrawn his fellowship; its professors believed he was spending too much time on public speaking rather than the study of Kant and Hegel.

"After a series of public speaking engagements, during my five years at Cornell University, I had become famous in the university. I noted in my diary that, in this small university town, I knew too many people and began to feel uncomfortable ... So I wanted to leave this small town of Ithaca and move to the big city New York, where there is an ocean of people. Each person walking on the street can keep to his own thoughts and not attract the attention of others," he said in his oral biography.

A third reason was that New York was one of the largest and most important cities in the U.S., where he could meet new people and different ways of thinking. A fourth reason was that Edith Williams had moved from Cornell to live in New York and study at Columbia.

Sources for Chapter Two

From Ithaca to Beijing: Hu Shih's Peripheral Centrality, by David Damrosch, Harvard University, 2016.

Cornell University website

Hu Shih and Female Emancipation in China, by Harriet Chien-ming Twanmoh, 1966, Australian National University.

The Creation of the Modern Japanese Language in Meiji-Era, by Paul H. Clark, University of Pittsburgh, Online Curriculum Project.

Young Hu Shih, by Tang Yan, Spring Hill Publishing Company, Taipei, 2020.

A Pragmatist and his Free Spirit: the Half-Century Romance of Hu Shi & Edith Clifford Williams, by Susan Chan Egan and Chou Chih-p'ing, the Chinese University Press of Hong Kong, 2009.

Oral autobiography of Hu Shi, as told to Tang De-gang, Yuanliu Publishing Company of Taiwan.

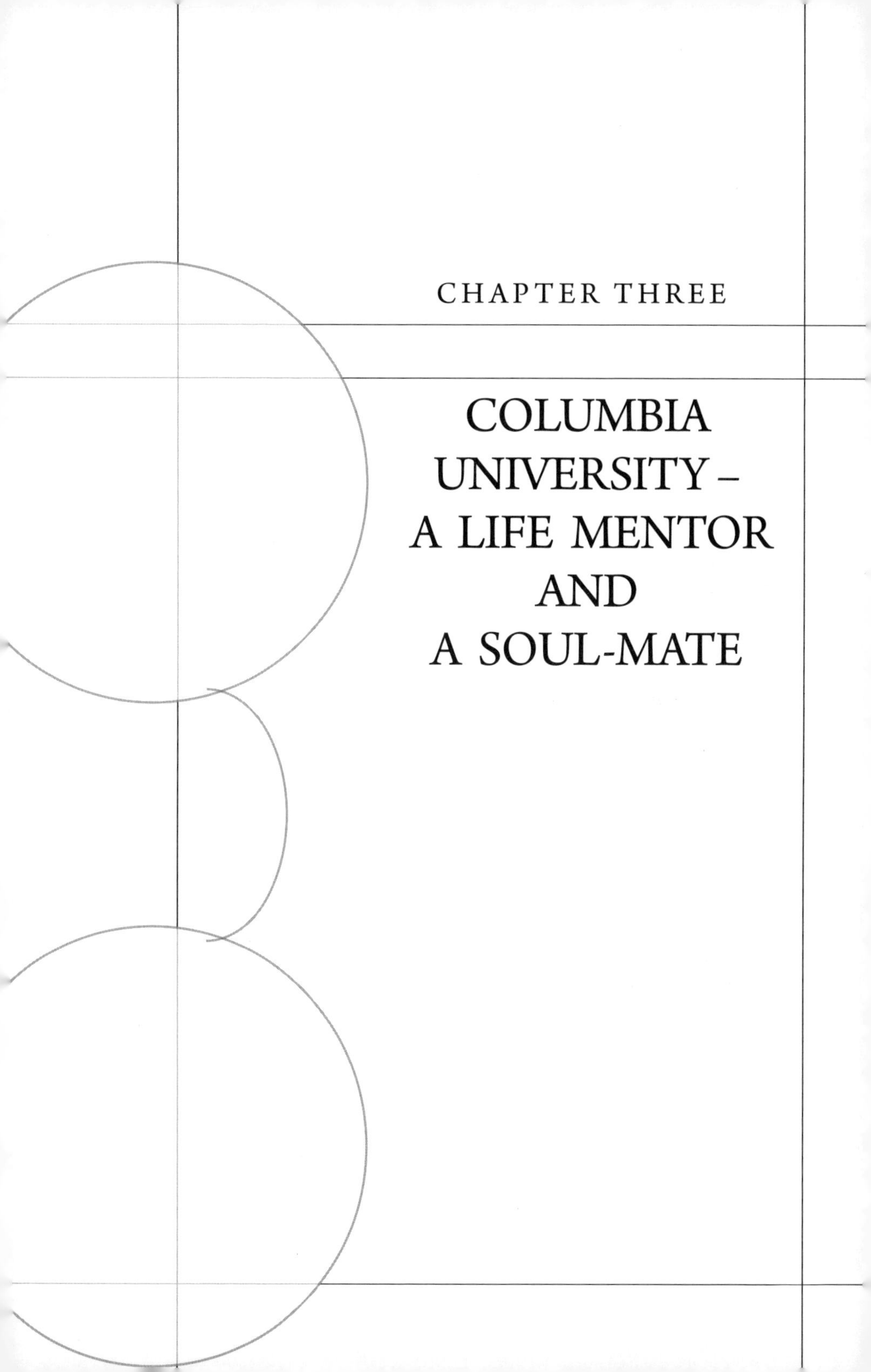

CHAPTER THREE

COLUMBIA UNIVERSITY – A LIFE MENTOR AND A SOUL-MATE

In the autumn of 1915, Hu moved to Columbia University in New York City, under the tuition of Professor John Dewey. Like Cornell, Columbia was one of the eight Ivy League colleges of the United States. It is a private institution founded in Manhattan in 1754; five of the Founding Fathers of the U.S. were its alumni.

In 1858, it established its School of Law. In 1896, the university moved to a spacious new campus in Morningside Heights; that was where Hu studied. In 1912, it established a School of Journalism, from the bequest of Joseph Pullitzer, after whom the famous prize is named.

By the time Hu arrived, Columbia had set up graduate faculties in political science, philosophy and pure science, making it one of the earliest centres in the United States for graduate education. Hu was moving from one centre of American excellence to another.

He was one of many of the elite of Republican China educated there; they became officials, scholars and business people. Among them were Wellington Koo, later his country's ambassador to Britain, the U.S. and France⊠and T.V. Song Zi-wen, a future Minister of Finance and Prime Minister, who studied there at the same time as Hu.

On his arrival in New York city on September 21, Hu moved into a room on the fifth floor of Furnald Hall, a residential building for Columbia students; it was the most modern hall of residents for male students. It overlooked Broadway, the 24-hour entertainment district – good for nightlife, but far from the peace and tranquility of Ithaca.

The building banned alcohol, gambling, pets and eating in the rooms; students could only meet women from 15:00 to 17:00 in a downstairs reception area. Hu lived in Furnald for the first of his two years at Columbia.

Hu Shih in the garden of Columbia University in 1916

In July 1916, for his second year, he moved into the apartment owned by the family of Edith Williams in a block on 92 Haven Avenue, close to 170th Street in Manhattan. It became vacant after she left New York in April 1916 to return to Ithaca to care for her sick father. A very pleasant apartment, it was quiet and overlooked the Hudson Road. Hu lived there with a fellow student, Lu Xi-rong, from Yunnan.

When he visited Ithaca, Hu used to stay with the Williams family. Edith remained there, caring for her father until his death in 1918. Her father was the only member of her family to take her art seriously. After his death, she gave up art and went back to her scientific training. From 1923 to 1946, she worked as the first full-time librarian of Cornell University's Flower Veterinary Library.

Friendship of 48 Years

After Hu's return to China in 1917, he had a busy and stressful life. So it is remarkable that he and Miss Williams continued their correspondence for more than 40 years. Her opinion and those of like-minded ladies Hu met in the U.S. strongly influenced his views on women. When he returned home, he saw the bondage in which millions of Chinese women lived, with bound feet, their marriages decided by their families and destined to serve their parents, their husbands and their children.

As on many subjects, he decided that drastic change was needed. During his lifetime, he wrote Miss Williams more than 300 letters and sent her flowers on her birthday, April 17. When he was travelling in Europe and America, he always sent her a postcard to describe his travels. Their friendship lasted 48 years. She kept all the letters. Three years after his death in 1962, Miss Williams donated them to Academia Sinica, the national academy of Taiwan, headquartered in Nangang, Taipei. It gave them to the Hu Shih Memorial Hall in Nangang, as part of its permanent collection.

As we noted in Chapter Two, Hu was careful to inform his mother of his friendship with Williams and put the two in touch with each other. He also told her, and everyone else, that he had a fiancée in China to whom he was engaged.

In a letter to Williams in March 1915, he wrote about his future wife: "I do not know what she thinks of me and of my ideas. She may have an 'idealised' notion of me in her own way. But she knows absolutely nothing of my ideas. For she can hardly write a short letter of greeting, nor can she read much ... I have long given up the idea of finding intellectual companionship in her ... I only know that I shall do all I can to make her happy – how successful I do not know. There was a time when I tried to urge her to acquire a better knowledge of reading and writing. But that was impossible – for numerous reasons. But I am an optimist. My mother can neither read nor write, but she is one of the best women that I know."

But, despite all this honesty, his family at home was anxious about his fidelity. This was not surprising. He had been away for five years in a society they could not imagine and mixing with attractive and eligible young ladies.

On August 28, 1915, Hu's mother sent him a letter, saying that she had heard rumours of his marriage. "The mother of Miss Jiang does not believe it but is feeling unwell and is anxious for her beloved daughter," she wrote. She asked him when he would return to China. In his reply, Hu said that he was already engaged to Miss Jiang and would not and should not break this engagement. "I consider myself an engaged man and that Miss Dongxiu will be my wife. When I meet other people, whether Chinese or American, I tell them that I am already engaged. Sometimes I tell you the names of lady friends. I have a clear conscience, I mean to be honest."

The mothers of Hu and Williams both would have strongly opposed a marriage between their children. This reflected the mainstream opinion in the United States and China regarding inter-racial marriage and mixed-race children; many states in the U.S. outlawed it. In addition, Williams was six years older than Hu.

If Hu had broken the engagement, it would be a humiliation for his fiancée and her mother . According to the traditions of Chinese society at that time, Miss Jiang would have been unable to marry another man. In the end, Hu accepted his mother's wishes. "I will follow Easterners in my family affairs, but in my ideas of society, the nation and politics, I will follow Westerners," he said. He told a friend: "We must live by the old conventions and marry the girl chosen for us. Ours is an intermediate generation which must be sacrificed both to our parents and to our children."

Other Chinese who studied in the U.S. faced a similar dilemma. The very first to graduate from an American university, Yale College, was Yung Wing in 1854. After he returned to China, he found that no Chinese parents would accept him as a son-in-law because they considered him too westernised. In 1876, he married Mary Kellogg, an American; such a union was rare at that time. It turned out to be a happy relationship; they had two sons.

Some Chinese of the late Qing dynasty period had, like Hu, marriages arranged for them by their families. Some chose to divorce this wife and marry a woman of their own choice; both Sun Yat-sen and Chiang Kai-shek did this. They continued to give financial support to their first wife. Like Hu, they lived in the two worlds of China – traditional and modern – and tried to balance the demands of both.

Reading Hu's curriculum vitae, you would have expected him to choose a "modern" wife, educated and with a background similar to his own: if not Miss Williams, then a Chinese lady. But he did not. The main reason was the love and respect he felt for his mother and the traditions of his home place. He could find from other people the intellectual companionship he needed.

Given his personality and that of Miss Williams, the two could have defied their families and convention and married. With his social and professional skills, he could have found a good job and comfortable life in the United States. Then China and the world would never have heard of Hu Shih.

Professor John Dewey, Guiding Light

The most important person Hu met during his seven years in the U.S. was John Dewey, his professor of philosophy at Columbia between 1915 and 1917. Dewey's ideas became a guiding light for Hu for the rest of his life.

In the first half of the 20th century, Dewey was one of the most prominent public intellectuals in the United States. When Hu arrived at Columbia, Dewey was 56 years old. He was professor of philosophy there from 1904 until his retirement in 1930. He wrote 40 books and published more than 700 articles in 140 journals. He advocated progressive education and liberalism, supporting such causes as women's suffrage. He was president of the teachers' union and sponsored the American Civil Liberties Union and National Association for the Advancement of Coloured People. During the inter-war years, he supported the Outlawry of War movement.

Born into a modest family in October 1859, Dewey attended the University of Vermont and Johns Hopkins University, where he obtained his Ph.D on the psychology of Immanuel Kant. After three years as a schoolteacher, he started teaching at the University of Michigan in 1884; it was the start of a lifetime career as a university teacher. In 1899, he was elected president of the American Psychological Association; in 1905, he became president of the American Philosophical Association. Some called him the Aristotle of his time.

This is what the Columbia University website says about Dewey: "His teaching style was characterised by long pauses and lots of backtracking, as if he was putting his ideas together as he spoke, the effect of which could either be inspiring or soporific. He also taught the philosophy of education at Teachers College, where his impact on educational theory and practice was both profound and controversial. With his wife, Alice, he helped establish laboratory schools, first at Chicago and later at Columbia. He said: 'The future of our civilisation depends upon the widening spread and deepening hold of the scientific cast of mind.'"

At Columbia, Hu took two of Dewey's courses, including "The Philosophy of Society and Politics".

What Hu found most useful in Dewey's encyclopaedic work, and most applicable to China, was his pragmatism or "experimentalism". When you face a problem, step back, research what it is and reflect what to do next: think of new ways of doing things, test and revise them: everything can be questioned. This inquiring spirit was a major reason for the rapid advance of the West since the industrial revolution – in medicine, science, industry, transport, armaments and other fields.

Throughout his life, Hu stressed this "experimentalism" –the application of scientific methods – to new areas, whether in politics and scholarship. He believed that social change should be gradual and evolutionary, that law was an instrument of politics and the individual has an important role in government and society. All this made him a political liberal and opponent of radicalism. In future, it would make him a strong opponent of Marxism.

The best way to change China, and other countries, he said, was through gradual and undramatic reform that solved specific problems. He advocated moderate and not extreme solutions. In a letter in 1916 to the father of Edith Williams, Hu wrote: "I do not condemn revolutions ... but I do not favour premature revolutions, because they are usually wasteful and therefore unfruitful ... My personal attitude is: 'Come what may, let us educate the people.'"

In his autobiography, Hu said: "Dewey teaches me how to think and consider the immediate problem in all cases, to regard all theories and ideals as hypotheses which are in need of verification, and to take into account the value of thought. These two men (Dewey and T.H. Huxley) make me understand the character and function of scientific method ... Only when we realise that there is no eternal, unchanging truth or absolute truth can we arouse in ourselves a sense of intellectual responsibility."

Thomas Henry Huxley was an English biologist and anthropologist whose philosophy was similar to that of Dewey. In 1889, Huxley wrote: "In matters of the intellect, follow your reason as far as it will take you, without regard to any other consideration ... In matters of the intellect, do not pretend that conclusions are certain which are not demonstrated or demonstrable."

In 1893, he wrote *Evolution and Ethics*, which was translated into Chinese and very influential in China; Hu read it in translation in Shanghai, before going to the United States. In the essay, Huxley argued that a man's emotions, intellect and preference to live in groups and raising children were part of his evolution and inherited. But, he said, values and ethics were not inherited. They were determined by a person himself and his culture.

Hu Shih with his life mentor, Professor John Dewey, during his time as ambassador to the U.S.

"Of moral purpose, I see not a trace in nature. That is an article of exclusively human manufacture," said Huxley. In other words, a person must learn morality. Dewey insisted that philosophy should not be in the abstract but deal with the problems of daily life and improve society.

"Its value lies in defining difficulties and suggesting methods for dealing with them," he said. From the two philosophers, Hu learnt to use the scientific method to solve problems and the need for society to promote morality and good behaviour. Hu was also impressed by Dewey's stress on the importance of popular education and civil society. Throughout his life, he expounded Dewey's ideas and called himself an "experimentalist".

With other students, Hu attended tea parties organised every month by Dewey's wife Alice at their home on the corner of Riverside Drive and 116th Street; the students met members of New York's art circle, including men with long hair and women with short hair.

Hu admired Alice Dewey, a highly educated woman who helped her husband set up an "Experimental School". She spent 10 years testing his theory of education. Later their eldest daughter, Evelyn, also studied education and travelled to different places to examine new educational methods for her father.

In October 1915, Hu witnessed a march of 40,000 people for women's suffrage down Fifth Avenue in Manhattan: both John and Alice Dewey took part. Hu was moved; he greatly admired this kind of family.

Legacy to Columbia

On December 12 1915, Yuan Shi-kai, who had become president of the Republic of China after Sun Yat-sen, declared himself emperor. He wanted to restore the hereditary monarchy from which the country had freed itself just four years before. It was a terrible blow to those trying to build a new, modern republic.

This was what Hu wrote to an American friend about Yuan: "I have come to hold that there is no short cut to political decency and efficiency ... Good government cannot be secured without certain necessary prerequisites ... Neither a monarchy nor a republic will save China without what I call the 'necessary prerequisites'. It is our business to provide for these necessary prerequisites – to 'create new causes'. "

Fortunately, Yuan's attempt to be emperor failed. Military leaders all over China opposed him; he abdicated on March 22, after just 101 days on the throne. He died of uremia on June 6, 1916, aged 56.

In 1919, after Hu's return to China, he invited Dewey to a lecture tour of the country; he served as host and interpreter. Dewey ended up staying for two years; it created a deep bond between him and the Chinese people. One of his daughters, Jane Dewey, said: "When he was in China, Dewey left feeling affection and admiration not only for the scholars with whom he had been intimately associated ... but for the Chinese people as a whole. China remained the country nearest his heart after his own."

During his life, Hu returned periodically to Columbia to teach and lecture. He assisted in a 1939 drive to increase the membership of the Alumni Federation. In 1960, he gave Columbia's East Asian Library a 25-volume set of his Chinese writings. Shortly after his death in 1962, the University established a graduate fellowship in his memory.

Creating a New Language

In studying Dewey's "experimentalism", Hu was always looking at what he could apply in China. In 1911, the Qing dynasty had been overthrown and the Republic of China established. Chinese were working hard to build a new, modern state; they were questioning which parts of their traditional society to keep and which to throw away.

This rare historical opportunity made Hu's experience in the United States especially precious. He was spending seven years in the country that had overtaken Britain as the world's largest industrial nation. Unlike China and the countries of Europe, it was born in 1776 without the burden of history, Emperors, Kings and an established church. It was able to experiment with new ideas and practise those which worked and discard those which did not.

The scientific spirit of Thomas Huxley and John Dewey was one reason for its astonishing progress – inquisitive, sceptical questioning and improving the life of ordinary people. Hu had the opportunity to learn about this new, dynamic country and ask what ideas and practices he could bring back and use in China.

One reason for America's success was mass education and mass literacy. Written with 26 Roman letters, English was easy to read and write. It was the instrument through which people learnt the tools of modernisation – from how to make automobiles to designing bridges and treating the sick. Chinese and Japanese students in the U.S. asked whether their languages were an instrument – or an obstacle – to literacy and modernisation.

A person learning Chinese had to memorise how to write a character, its sound and its meaning; it was far more difficult than learning a language with an alphabet. Those learning Japanese also had to learn Chinese characters, but only a limited number; the rest of the language was written into two alphabets, hiragana and katakana.

Once you had learnt these two alphabets, you could read the words; it was easier than learning Mandarin. Some Chinese proposed abandoning the characters completely and replacing them with words written in Roman letters. But such a plan has never been feasible because many characters have the same sound; if written only in Roman letters, their meaning would be unclear and ambiguous.

The first romanisation of Chinese was devised by two Jesuit missionaries in Beijing, Matteo Ricci and Michele Ruggieri, in the 1580s, for their Portuguese-Chinese dictionary. Over the next three centuries, foreign missionaries and diplomats developed other systems. In Taiwan in the 19th century, Presbyterian missionaries invented a romanisation system for the Taiwanese dialect. They did so after they gave their congregations the Bible in Chinese – but discovered that most could not read the characters; this system is still in use in Taiwan today. Chinese scholars also invented romanisation systems. But none of these systems was widely used within China; they were mostly to help foreigners learn Mandarin.

Since 1982, the international standard for writing Chinese in Roman letters has been Hanyu Pinyin. This was created between 1955 and 1958 by scholars in the People's Republic of China (PRC), to help school children learn Mandarin and reduce widespread illiteracy.

In 1958, the government introduced it in primary schools. A student learns both a character and its pronunciation in Hanyu Pinyin; for example, he learns three hundred as " 三百 , san bai" and goat as " 山羊 , shanyang". Since then, Hanyu Pinyin has helped tens of millions of Chinese children, as well as foreigners, learn the language.

The chief of the team that developed it was an extraordinary man named Zhou Youguang. He was born on January 13, 1906, son of a Qing dynasty official; as a student, he wrote classical Chinese. During World War Two, he worked for a Chinese bank in Chongqing, the wartime capital.

After the war, the bank sent him to New York and London; he returned to China in 1949. During the Cultural Revolution, he was sent to the countryside to be "re-educated" and spent two years in a labour camp. After 1980, he worked with two others on translating *Encyclopedia*

Britannica into Chinese.

After he was 100, he published 10 books; some are banned in China. He advocated political reform and democracy and criticised the attacks of the Communist Party on traditional culture after it took power. He died at his home in Beijing on January 13, 2017, one day after his 111th birthday. When people praised him as the "father of pinyin", he replied modestly that he was, rather, "the son of pinyin".

But, for Hu and his fellow students, no such sophisticated system of romanisation was available. In their rooms and meeting halls, they continued the intense debate on the future of the language they had started at Cornell University.

Hu strongly supported the use of vernacular Chinese as the standard written form. In the *Chinese Students' Monthly* of June 1916, he wrote an essay on "The Teaching of Chinese as It Is". The journal was published by The Chinese Students' Alliance in the U.S. and distributed from Ithaca. Here are excerpts from the essay:

"The teaching of Chinese constitutes a far more urgent problem, because it is the language which records our past and present civilization, which is the only means of inter-provincial communication, and which is the only available instrument of national education. There are a few generalisations which I consider to be of great importance in discussing the problem of teaching Chinese as it is. The first of these is that what we call our literary language is an almost entirely dead language. Dead it is, because it is no longer spoken by the people. It is like Latin in Mediaeval Europe; in fact, it is more dead (if mortality admits of a comparative degree), than Latin, because Latin is still capable of being spoken and understood, while literary Chinese is no longer auditorily intelligible even among the scholarly class except when the phrases are familiar, or when the listener has already some idea as to what the speaker is going to say.

"The second generalisation is that we must free ourselves from the traditional view that the spoken words and the spoken syntax are 'vulgar'. The Chinese word vulgar means simply 'customary' and implies no intrinsic vulgarity. As a matter of fact, many of the words and phrases of our daily use are extremely expressive and therefore beautiful. The criterion for judging words and expressions should be their vitality and adequacy of expression, not their conformity to orthodox standards. The spoken language of our people is a living language: it represents the daily needs of the people, is intrinsically beautiful, and possesses every possibility of producing a great and living literature as is shown in our great novels written in the vulgate."

Hu decided that, from the beginning of July 1916, he would no longer write poems in classical Chinese, only in the vernacular.

Many of Hu's fellow students disagreed strongly with him. His sharpest critic was Mei Guang-di, who had also gone to the U.S. in 1911 under the Boxer Indemnity Scholarship scheme. A year older than Hu, Mei was also a native of Anhui and extremely intelligent. In 1904, at the age of 12, he passed the county-level imperial exam of Xiu Cai (秀才), the first step to a career in the civil service. The next year the Qing government abolished the imperial exam system; so Mei could not proceed on the career he had chosen. He and Hu met in Shanghai when they were studying in high school there.

In the U.S., Mei first studied History and Politics at the University of Wisconsin, then History and English Literature at Northwestern University. He went on to do doctorate studies in English literature at

Harvard University. He was not as impressed as Hu by the United States; he saw political incompetence, social unrest, intellectual confusion and decadence. His view was that, while it was more advanced than China in material terms, it was behind in terms of morality. He wanted to revitalise Confucianism to save China.

In 1915, with three other Chinese students, Mei went to Ithaca to spend the summer. The four discussed language reform with Hu. The discussions became heated and tense, with Hu and Mei on the opposite sides. "Mei absolutely did not recognise that classical Chinese was a half-dead or completely dead language," wrote Hu. "Because he refuted me, so I had to think in detail of my own position. The more he proposed preserving the old, so I gradually became more intense. Then I often advocated the idea that Chinese literature needed a revolution. The slogan of 'a literary revolution' was born of our wide-ranging discussions that summer."

In a poem addressed to Mei on September 17 that year, Hu first used the phrase "literary revolution". Mei strongly supported the retention of classical Chinese as the written language. For him, the answer was not revolution but reform of the ancient learning. The solution was for people to master Chinese by reading ancient books and literature.

He said that Hu was too influenced by the vulgar poetic movements of the West. Over the next months, while Hu was studying at Columbia, the two men debated the issue intensely by mail. Mei's opinion reflected that of many scholars; they believed the classical language was the heritage of Chinese knowledge and wisdom over many centuries. What qualifications had this opinionated young man to propose abandoning this language?

At university, Hu had not specialised in Chinese language and literature – and he was studying not in China but abroad, using English. Mei continued to write classical Chinese into the 1940s, two decades after the vernacular had become standard. Throughout his life, he refused to write the vernacular version. He died in Guiyang in December 1945.

Revolution of the Language

During his two years of graduate study at Columbia, Hu devoted much time to the study of language reform. His doctorate dissertation was "The Development of the Logical Method in Ancient China," including "pragmatic tendencies" in early Chinese thought. Later he expanded and rewrote this in Chinese as the first volume of *An Outline of the History of Chinese Philosophy*.

Before he returned to China, Hu wanted to present his ideas on language reform to the audience at home. He did this through an article published in Beijing in *New Youth*. This was a magazine started by Chen Duxiu , with its first issue on September 15, 1915.

Like Hu, Chen was a native of Anhui, born in Anqing in October 1879. He studied in Japan from 1902 to 1908, where he campaigned against the Qing government and the interference of foreign powers in China. After his return from Japan, he started *New Youth*. It was the magazine for its time in history. It quickly became the most popular and widely distributed periodical among the intelligentsia in China; like Hu, people were searching for the best way to use the new liberties the revolution had given them.

The magazine's articles criticised traditional thinking and customs and advocated science, democracy and human rights. It opposed the hypocrisy of men being allowed to have several concubines, while women were expected to be faithful to their husband for their lifetime. It advocated marriage through free choice of the partners, not arranged by the parents: and the smaller families common in the west instead of the extended ones common in China. It expressed many of Hu's ideas; this made it a natural platform for him to promote the 'new language'. He wrote the article in the New York apartment of Edith Williams. It appeared in the January 1917 issue, with the title "A Preliminary Discussion of Literature Reform."

It proposed eight guidelines for effective writing. The first was "to write with substance ... Feeling is the soul of literature. Literature without feeling is like a man without a soul." It must contain thought. "Thought does not necessarily depend on literature for transmission, but literature becomes more valuable if it contains thought ... In recent years, literary men have satisfied themselves with tones, rhythms, words and phrases, and have neither lofty thoughts nor genuine feeling. This is the chief cause of the deterioration of literature."

The second was not to follow ancient writers. "Literature changes with time. Each period from Zhou and Qin to Song, Yuan and Ming has its own literature ... Each period has changed in accordance with its situation and circumstance, each with its own characteristic merits ... I have always held that colloquial stories alone in modern Chinese literature can proudly be compared with the first-class literature of the world. Because they do not imitate the past but only described the society of the day, they have become genuine literature."

The third was to respect grammar. The fourth was to reject melancholy. "Nowadays young writers often choose names like 'Cold Ash', 'No Birth" and 'Dead Ash' as pen names and, in their prose and poetry, they think of declining years when they face the setting sun, and of destitution when they meet the autumn wind." He wanted writers to be more positive and optimistic.

The fifth was remove old clichés. "Writers should describe in their own words what they personally experience. So long as they achieve the goal of describing things and expressing the mood without sacrificing realism, that is literary achievement."

The sixth was not to use allusions. By this, Hu was referring to the practice of comparing present and historical events even when there was no meaningful analogy.

The seventh was not to use couplets and matching words. "We must not waste our useful energy on the non-essentials of subtlety and delicacy. This is why I advocate giving up couplets and rhymes." The eighth was not to avoid popular expressions.

"From the modern point of view, the Yuan period should be considered as a high point of literary development; unquestionably, it produced the greatest number of immortal works. At that time, writing and colloquial speech were the closest to each other, and the latter almost became the language of literature. Had the tendency not been checked, living literature would have emerged in China, and the great work of Dante and Luther (who inaugurated the substitution of a living language for dead Latin) would have taken place in China. Unfortunately, the tendency was checked in the Ming when the government selected officials on the basis of the rigid 'eight-legged' prose style ... Thus the once-in-a-millennium chance of uniting writing and speech was killed prematurely, midway in the process.

"But, from the modern viewpoint of historical evolution, we can definitely say that the colloquial literature is the main line of Chinese literature and that it should be the medium employed in the literature of the future. We should use popular expressions and words in prose and poetry. Rather than using dead expressions of 3,000 years ago, it is better to employ living expressions of the 20th century. Rather than using the language of the Qin, Han and the Six Dynasties, which cannot reach many people and cannot be universally understood, it is better to use the language of the *Water Margin* and *Journey to the West*), which is understood in every household."

This was a remarkable essay by a graduate student just 26 years old, especially one who had spent the last seven years listening to lectures and discussions about European writers and philosophers. It was only with his fellow Chinese students that he could debate the issue of vernacular language and refine his ideas on the subject. It also tells us that, between writing his doctoral dissertation, busy social life and long letters and talks with Miss Williams, he found the time to read the Chinese books he had brought with him.

The essay challenged the mainstream of Chinese scholarship, that the classical language was the proper form for literature. It was a sign of his intellectual learning and self-confidence and, many said, his ambition and arrogance. "This article caused an enormous reaction in Chinese cultural circles," said Hu . Chen Duxiu, editor of the magazine, strongly supported it. The issue became a subject of national debate among China's intellectuals. Many sent letters to the magazine. By the time he returned home, Hu was well known.

What was new about Hu's essay? Since the 1890s, Chinese intellectuals had debated the wide gap between the written and spoken languages and what to do about it. Authors had written books, novels and articles in the vernacular, especially when they wanted to reach a wide audience.

The consensus among the intelligentsia was that classical Chinese was the standard, and the vernacular was supplementary, like a simpler version of the proper language. No, said Hu, the vernacular should replace classical and become the standard Chinese for books, magazines, education and daily life. That is why he – correctly – called it a "revolution". He wanted to change the written form of the world's oldest written language. In this, he was the pioneer; it was his greatest, and most long-lasting, contribution to Chinese civilization. It has helped to bring literacy to millions of people. More than a century later, the written and spoken Chinese used in the PRC, Taiwan, Hong Kong and around the world is the one he advocated.

Going Home

Hu Shih's writings attracted the attention of many people, including Cai Yuan-pei, president of Beijing University. Chen Duxiu recommended that he hire Hu as a professor at the university. Cai took his advice and invited Hu to join the faculty of liberal arts; Hu was delighted. This was China's most famous university and Cai was one of the country's leading educational reformers; it was an honour for someone who had only just graduated.

Cai himself had had an extraordinary journey. Born in Shaoxing, Zhejiang in 1868, he studied classical Chinese and, in 1892, entered the Hanlin Academy, the highest honour for a scholar in imperial China.

From 1907 to 1911, he learnt German in Berlin and then studied philosophy and ethics at the University of Leipzig. In 1912, he served as first Ministry of Education of the new Republic of China. From 1913 to 1917, he studied in France and helped to educate Chinese workers helping the Allied countries during World War One. In late 1916, he returned to China to become president of Beijing University on January 4. We will learn more about him in the next chapter.

On May 4 1917, Hu handed in his doctorate, which had taken nine months to write. The title was "A Study of the Development of Logical Method in Ancient China". Before he left for China in June, he stayed in Ithaca with Miss Williams and her family.

He wrote in his diary: "It is difficult for me to leave Ithaca. Mrs Williams and Miss Williams have treated me as one of the family and it's especially hard to leave them. I often said that my home is where my friends are. Most of my friends are this country. On leaving this home of my own making for the home of my fathers, I am not sure whether I'm more happy or sad."

After crossing North America by train, he took a boat from Vancouver on June 13, with four Chinese friends. After stops in Yokohama and Nagasaki, he arrived in Shanghai on July 10. During the six-week journey home, he wrote seven letters and one post card to Miss Williams. His body and his mind were returning to China – but he had left his heart behind in Ithaca.

There can be few foreign students in the United States who have learnt and achieved so much as Hu. First, he had to master the foreign language he had to use for talking, reading, writing academic papers and giving public speeches. He devoted much time to his academic studies. He also threw his energy into social and intellectual life, in student societies and magazines, and developed a wide network of friends among fellow Chinese students, foreigners and Americans, especially his professors and their families. He would maintain contact with some of them for many years.

He was able to do all this thanks to his fluency in English, self-confidence, outgoing personality and ability to make friends. He delivered many public speeches, improving his skill in how to communicate to an American audience. He learnt more about American life, thinking and culture than most Chinese students.

This was a remarkable achievement. A student from Europe in the United States found a common history, tradition and religion and those from Britain a common language. All this made it easier for them to adapt to American life and society. But those from Asia had to cross a huge divide; everything was new and strange – dress, language, religion, social manners and ways of thinking. They had to learn everything from scratch. To achieve all he did during his seven years, Hu had to call on great intellectual, social and spiritual energy; only by living in this foreign environment did he realise that he had it. His English became fluent.

For the rest of his life, he was as comfortable living in the U.S. as in China; he was at ease with the elite of American society he met on both sides of the Pacific. There were few people in China of which that could be said. He would spend 21 of his 71 years living in the United States. This initial stay also changed his outlook on life. He wrote that he was most impressed by the "naïve optimism and cheerfulness of the American people and that, as a result, he came to believe that "in this land, there seemed to be nothing which could not be achieved by human intelligence and effort."

Sources for Chapter Three

Young Hu Shih 1891-1917, by Tang Yan, Spring Hill Publishing Company, Taipei city, first edition June 2020.

Columbia University website

A Pragmatist and his Free Spirit, the Half-Century Romance of Hu Shih and Edith Clifford Williams, by Susan Chan Egan and Chou Chih-p'ing, Chinese University Press, Hong Kong, 2009.

If Not Me, Then Who? Hu Shih Volume One, Jiang Yong-zhen, Linking Publishing Company, Taiwan, first edition January 2011.

Hu Shih and Chinese language reform, by Victor Mair, in China Heritage Annual 2020.

Hu Shih Oral Autobiography, as told to Tang De-gang, Yuanliu Publishing Company, Taiwan, published in November 2010.

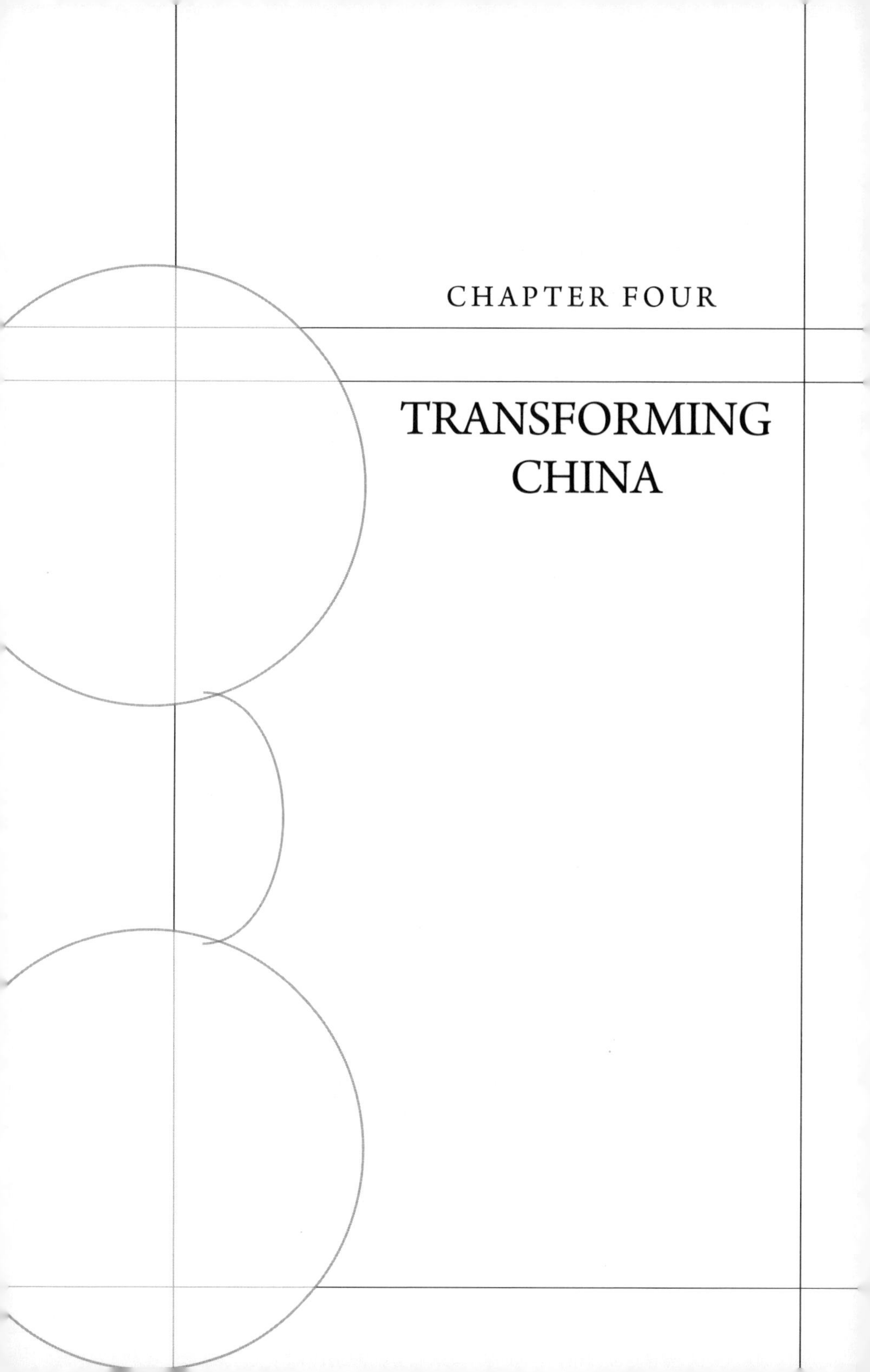

CHAPTER FOUR

TRANSFORMING CHINA

Hu returned to China in June 1917. The next six years were intense and dramatic. He married the wife chosen by his mother and they had three children. Hu began his teaching career at Beijing University. His proposals to adopt the vernacular as the national language were widely accepted, despite strong opposition.

To spread new knowledge to a wide audience, Hu organised nationwide tours by famous foreign scholars. He promoted ideas he had learnt in the U.S., including: the right to choose your partner in marriage; birth control; sexual equality between men and women; and the science of his mentor, John Dewey, to treat everything with scepticism and demand proof.

His ideas reached a nationwide audience through his lectures and the new media of newspapers and magazines written in the vernacular. This young man, not yet 30 years old, became a national figure and influenced thousands of people.

Marriage "Keen Curiosity"

After Hu's arrival, he went first to report to Beijing University (BU), where Cai Yuan-pei had given him a teaching post. His writings from the U.S. had already earned him a reputation among the intelligentsia.

Among his admirers was Lin Yu-tang, then a student at Tsinghua University in Beijing. He wrote: "Hu Shih returned with national acclaim to join Beijing University, and I was at Tsinghua to greet him. It was an electrifying experience."

The two men had much in common. Lin studied in the United States, as well as in Germany, and taught English Literature at Beijing University.

Lin became a prolific author, in Chinese and English. His translations of classic Chinese texts were bestsellers in the West.

Hu's second duty was to return home to Jixi, which he had not seen for seven years. He arrived in mid-August and stayed for a month. His mother was in poor health but as loving as ever.

He walked 16 kilometres to the home of his fiancée Miss Jiang Dongxiu. Hu made it as far as her bedroom – but she was hiding in bed, with the curtains drawn. She did not want to see him – he had to settle for a meeting with her brother and other members of the family. Was this a protest for the 14 years he had kept her waiting?

In preparation for marriage, she had unbound her feet and began to learn how to read and write. On September 13, Hu left for Beijing. During his long absence, Miss Jiang had spent several months every year with his mother. In January 1916, Miss Jiang's mother died; her final wish was left unfulfilled – to see her daughter married.

She came from a wealthy, educated family with land and servants; it had produced many scholars and senior officials. Under normal circumstances, the two would have married much sooner, but the event was put off by Hu's studies abroad.

In a letter to Miss Williams on November 21, he wrote: "I cannot say that I look forward to our wedding with gladness. It is with a sense of keen curiosity that I approach the eve of a great experiment – the experiment of living!"

After negotiations between the families, the two were finally married on December 30, 1917; Hu was 26 and his wife one year older. She wore a

Hu Shih with his wife Jiang Dongxiu

black satin outfit that he had brought from Beijing and he a Western suit. She was carried in a decorated sedan chair over the mountain from her home and greeted by firecrackers. The two saw each other in person for the first time – previously they had seen only photographs.

They visited the Hu family temple, bowing three times to the ancestral tablets, and attended a lavish wedding banquet. They became intimate immediately. The two had a honeymoon of one month before Hu returned to Beijing. Miss Jiang stayed behind to look after his mother, who had fallen ill shortly after the wedding, probably from the stress of organising it.

In a letter on February 19, 1918 to Miss Williams and her mother, Hu said that he was happy to report that he and his wife were quite happy and believed that they could get on well together. One thing that annoyed him was her refusal to accept education and write letters regularly to him. During his seven years in the U.S., she had sent only a few.

It is full of irony that a leading pioneer of China's modernisation who had argued against arranged marriages accepted one himself. He did it out of respect for the mother who brought him up and to whom he owed so much. "If I had repudiated the engagement, several people would suffer the rest of their lives and the pain I would endure from my conscience would hurt me more than anything else," he said in 1921 in a conversation with a friend which he recorded in his diary.

In the summer of 1918, Miss Jiang moved from Jixi to Beijing to join her husband. She unbound her feet – but they never regained their normal shape and function. Like Hu's mother, she began her education after her marriage. Her letters to her husband were full of grammatical errors. Her limited education made it difficult for her to follow current events and understand her husband's writings and thoughts.

The couple lived in two different worlds. She gave him two sons and a daughter. She was an excellent cook, especially of Anhui cuisine. While she shared little in common with her husband, she had a self-confidence that came from a happy upbringing in a well-established family. She enjoyed mahjong with her friends. On his 32nd birthday, she gave him a "stop drinking" ring, which he wore on his finger. When friends pushed him to down another glass, he pointed to the ring and said it was the order of his wife.

In November 1918, less than a year after the wedding, Hu's mother died, aged just 46. He was heart-broken. In a letter to Miss Williams in March 1919, Hu wrote: "My mother – whom you knew well through what I have told you – died of influenza last November! The loss was too great for me to bear. She was only 46 years old when she died. She had suffered everything for over twenty years merely for my sake, and I have never been able to offer the little happiness which I am only beginning to be in a position to give! After my wedding, I left my wife to be with her. But she knew that I was working hard, so she sent my wife to Beijing to be with me. She was very happy when she learnt that we were soon to have a child. But she never lived to see her grandchild. When influenza came, she never allowed anybody to write me of her illness ... My sole consolation is that I have been able to see my mother when I returned from America after 11 years of absence from her."

After they received news of her passing, Hu and his wife rushed home. They buried her on December 17, 1918, Hu's birthday. He was the more grief-stricken because he had not gone to see her during the last months of life; her only child, he was not at the bedside of his mother during her final hours on earth.

There was less news from his soul mate across the Pacific Ocean. For three and a half years after Hu's return to China, Miss William did not write to him. Was it out of respect for his new wife and not wishing to come between them? Or because of her complicated feelings toward him?

Beijing University and the New Language

The Qing government had been slow to establish universities, a basic building block of the modern state. In 1877, neighbour Japan established Tokyo University, the first of nine imperial universities which are considered the Ivy League of Japan. They trained the elite that soon made the country the leading industrial and military power in East Asia.

It was only after its defeat by Japan in 1894-95 that the Qing was persuaded to set up a university in Beijing. The Imperial University of opened on December 31, 1898, with 150 students and an American Presbyterian missionary as its first president. In 1904, it sent 47 students overseas, the first time a Chinese university had sent students to study abroad.

After the Xinhai revolution, it was renamed Government University of Peking. On December 26, 1916, President Li Yuan-hong appointed Cai Yuan-pei as president of BU. Before Cai's arrival, it had fewer than 1,000 students. Accounts of the time describe it as very bureaucratic; many students were wealthy and more interested in mahjong, eating and drinking than study.

Cai transformed the institution. He brought ideals and practices he had learnt at the University of Leipzig, where he had studied philosophy, psychology and art history. One of his professors was Karl Lamprecht, who founded at the university the Institut für Kultur und Universalgeschichte, The Institute for the Study of Comparative World and Cultural History; Lamprecht wrote a 13-volume history of Germany.

As with Hu's experience at Cornell and Columbia, the years of study in Europe had opened Cai's eyes to the breadth and intellectual rigour of

Western history and scholarship. He took it upon himself to transform BU into China's most important institute of higher learning. He expanded the number of faculties to 14; he increased the enrolment to more than 2,000 students from 900 at the end of 1914.

The faculties were Chinese, English, French, German and Russian literature, mathematics, physics, chemistry, geology, philosophy and history, economics, political science and law. Later he added education, Oriental studies, biology and psychology and extended the undergraduate programme from three to four years.

He set up learned societies where faculty and students could discuss many subjects and he promoted freedom of thought and tolerance for the opinion of others, critical thinking and the introduction of Western ideas. Cai wanted an environment where people would speak their minds and an education outside the control of politicians. In the conservative world of Chinese education, this was a revolution.

To introduce such new ideas, Cai needed new professors to teach them. For this reason, he hired Hu Shih, as well as Chen Duxiu and Li Dazhao. Both had studied in Japan and went on to become founder members of the Chinese Communist Party in July 1921.

It was only during a period of sweeping change that Hu could obtain such a prestigious post at the age of just 26. He earned the considerable salary of 280 yuan a month. In 1918, in preparation for his wife's arrival, he rented a traditional courtyard home near the university with 17 rooms. It had no bathroom or running water; to wash himself, Hu had to go to a nearby public bathhouse. Fortunately, he adapted as easily to the inconveniences of living in an ancient Chinese city as he did to the comforts of Ithaca and New York.

Hu Shih in his office in Beijing University

Also working at BU at that time was Mao Tsetung; he earned eight yuan a month as an assistant in the library. He went to hear a lecture by Hu; but, because he was not an enrolled student, he was not admitted. His uncouth manners and thick Hunan accent did not endear him to the sophisticated people who worked in the university. Many believe the impolite treatment he received there was one reason for his hatred of intellectuals and the deadly campaigns he unleashed against them after he took power.

Hu began his teaching career with a course on ancient Chinese philosophy. Many students were uncertain of this professor, who was so young and had just spent seven years in the U.S. Discarding what had been taught before, Hu wrote his own material which he handed out to the students. The course began in the ninth century B.C.; he considered material before that as unreliable.

His lectures and teaching methods gripped his students. He was modest and approachable, unlike many other students who had returned from abroad; they considered themselves superior and looked down on others. Hu turned the course material into a book, *An Outline of the History of Chinese Philosophy, Volume I,* published in 1922. It was the first book of its kind to be written in the vernacular and using Western methodology. In February 1919, the Commercial Press in Shanghai published the book. It was so popular that, within three years, it went through seven editions.

Hu also taught History of Western Philosophy, a Brief History of Chinese literature, English Poetry, Famous European Literature and an Intellectual History of the Tang and Song dynasties.

Hu was fortunate to work under someone as enlightened as Cai, who provided an environment perfectly suited to Hu's active, restless mind. A traditional, conservative Chinese university would not have hired him.

Hu thought highly of President Cai. "He was a great leader, interested in the 'literature revolution'. He studied philosophy in Germany for a period and so was a scholar able to accept practices of the new era. He was a modern person very able to accept new ideas and opinions." Cai also appointed Hu to the university's appraisal committee; he was able to offer his opinion on many issues facing the institution. He also joined the editorial team of the *New Youth* magazine.

Hu continued to promote the vernacular language. In 1918, he wrote an article "On the Theory of Constructing the Literary Revolution". He argued for the use of vernacular in literature and this literature as the national language. His ideas were accepted with astonishing speed; more and more people wrote articles and books in the vernacular.

One major reason for its appeal was that this was everyday language. A person could compose something from his own knowledge and experience, without having to devote years of study to master classical Chinese; and someone with a basic knowledge of characters could read it.

In 1919 and 1920, more than 400 primary schools published their own magazines using the vernacular, using printing or oil presses. In 1920, the Ministry of Education issued an order to primary schools across the country that all their materials for first and second year students should use the vernacular. "This was beyond my expectation. We only used four years, arguing that schools should use the vernacular and not classical Chinese. It was a complete success."

After 1922, all the material in primary schools used vernacular. From 1919, newspapers and magazines began to use it, making their content available to a much larger audience. Many authors used it to write novels. The speed of the change from classical to vernacular was dramatic. It was as if a large volume of water had built up behind a dam. The birth of the republic, the end of the established order and the persuasive arguments of Hu and other reformers broke down the dam – and the water poured through.

But many scholars questioned Hu's qualifications to propose sweeping proposals about China's language. He had been educated in Chinese only until the age of 17. His university studies were in English; his doctoral thesis was in English. He was young; his knowledge of Chinese language and literature could not compare with those who had devoted their lives to the study of them.

They also criticised the quality of his poems and other creative work, saying that it was below that of other authors, like Lu Xun. Just as Latin was the basis of many European languages, so classical Chinese was the parent of the vernacular, they argued. The two versions were not opposed to each other, but complementary.

Firestorm from France

During his seven years in the United States, Hu had been shielded from the convulsions that shook China and Europe. In China, it was the Xinhai revolution of 1911 and Japan's seizure of the German concession of Qingdao, Shandong province in the autumn of 1914. Japan followed this with its 21 Demands on the Beijing government in January 1915.

In Europe, World War One broke out in July 1914; it became the most devastating conflict in history. The United States and China remained neutral until April and August 1917 respectively, when each declared war on Germany. Like his fellow Chinese students in the U.S., Hu followed the progress of the war with great attention; but they were at a safe distance. Then, in May 1919, the war arrived at his doorstep in Beijing with a bang.

After Germany's surrender in November 1918, the victorious powers convened a peace conference at the Versailles Palace outside Paris to divide the spoils. They first concentrated on the main battleground, Europe, then Africa and the Middle East and, finally, Asia, where fighting had been very limited.

The main issue was the German concession in Shandong; should the victors award it to Japan, which had occupied it in 1914, or return it to China? Both countries were on the winning side; their delegations sat around the same table in the cavernous negotiating hall in Versailles.

China had contributed 140,000 labourers to the Allied war effort in France and Belgium; they had worked close to the front lines but were not in combat.

The Japanese contribution was, from April 1917 until the end of the war, a naval squadron of nine ships. It took part in 348 missions to escort Allied troopships in the Mediterranean; they accompanied 750 ships, covered over 240,000 nautical miles and lost 72 sailors. The ships saved hundreds, perhaps thousands, of lives of Allied soldiers in ships; otherwise, they might have been sunk by German and Austrian submarines in the Mediterranean.

Britain, France and the U.S. compared the contributions of China and Japan to the war effort; in addition, Japan was the most important ally of Imperial Britain in the Far East. In April 1919, the three victors decided to award the concession to Japan, not China. If it signed the Peace Treaty that ended the war, Beijing would have to accept these terms. Some in the government were willing to do so.

When the news reached Beijing, it was greeted with anger and disbelief. Was this the reward for the blood and sweat of the 140,000 Chinese who had carried munitions, dug trenches and cleared unexploded bombs in the fields of Picardy and Flanders? Of them, about 10,000 had died from shelling, landmines, disease and Spanish influenza.

The China of 1919 was different to the country of 1895, after the humiliating defeat by Japan. The 1911 revolution and the social awakening it provoked had created a civil society that had never existed before. This new sense of national pride was especially strong among students. In part, this was the harvest of the seeds sown by Cai, Hu and their fellow teachers at Beijing and other universities.

On the afternoon of May 4, 3,000 students, many from BU and carrying nationalist banners, gathered in Tiananmen Square to protest in the quarter which housed foreign embassies. The police barred them from entering. So the students broke into the house of Cao Ru-lin, Minister of Transport; he supported ratification of the treaty. They beat up some of his family and friends and set the house on fire. Police arrested 32 students and put them in jail. This sparked strikes and demonstrations in other cities and a nationwide boycott of Japanese goods; police arrested more students, and tens of thousands poured into the streets.

In Shanghai, centre of China's economy, merchants and workers launched a general strike. This was the first mass popular protest in Chinese history; it was called the May 4 Movement. The students established Tiananmen square as a centre of protest for the next 100 years.

After the government put the students in prison, Cai Yuan-pei resigned as president of BU in protest. Such was the intensity of public anger that, on June 12, the cabinet of Prime Minister Duan Qi-rui resigned. The government fired three pro-Japanese members of the Cabinet and told its delegation at Versailles not to sign the Peace Treaty.

One member of the delegation was Wellington Koo; like Hu, he had earned a Ph.D at Columbia University. He delivered an eloquent speech in English at the Versailles talks, pleading his country's case. China was the only country at Versailles not to sign the treaty on June 28. It later signed a separate agreement with Germany. It was the first time in Chinese history that a popular movement had changed government policy.

In September, after the government released the students, Cai returned to his post at BU. On February 4 1922, China signed a treaty with Japan; it agreed to hand back the Shandong concession and the railway that ran through it. China achieved this without the threat or use of military force; it was one of the first diplomatic victories of the young Republic. The popular mobilisation and awakening of the May 4 Movement had profound consequences for China in society, culture and politics. One of the most important was the establishment of the Communist Party in July 1921, which we will describe in the next chapter.

This is what Hu wrote about the dramatic changes of 1919, in an article in December that year for the *Chinese Social and Political Science Review*: "The events of 1919 gave us a new lesson. It was the non-political forces – the students, the merchants, the demonstrations and street orations and the boycott – that did the work and triumphed. This was a great revelation and produced a new optimism."

For him, political reform was only a small part of China's renaissance. "We still have the masses to educate, the women to emancipate, the schools to reform, the home industries to develop, the family system to reshape, the dead and antiquated ideas to combat, the false and harmful idols to dethrone ... The scientific spirit is beginning to make itself felt in the Chinese intellectual world today. It first shows itself in the attitude of doubt. The question 'why' is heard everywhere. Why should we believe in this or that idea. Why should this or that institution still exist today?"

Barnstorming China

On February 9 1919, Hu's former teacher Professor John Dewey arrived in Japan for a series of lectures on philosophy at Tokyo Imperial University. After the lectures, he planned to spend six weeks touring China before returning home.

On hearing this news, Hu discussed with colleagues at other Chinese institutions how to raise funds to cover the costs of Dewey lecturing in China. Cai Yuan-pei wrote to the president of Columbia University; he gave his approval to Dewey teaching at BU for a year.

Dewey and his wife were touched by the invitation and eager to visit China; but they were unsure how long to stay and whether their Chinese hosts would be able to pay them. They decided to take one step at a time. On April 30, the two Deweys arrived in Shanghai. They were met at the pier by Hu and other former students; they escorted the couple to a guest house.

After several days touring around Shanghai, Dewey gave his first lectures on May 3 and 4 at the Jiangsu Education Association in Nanjing. He spent two more weeks lecturing in Nanjing, before moving to Beijing. Hu served as his translator.

The May 4 movement broke out four days after Dewey's arrival. A dedicated social activist, Dewey was energised by everything he saw; he was witnessing history in the making. He saw protests and demonstrations in Shanghai, Beijing and other cities. A tourist visit of six weeks turned into a stay of 26 months. Dewey gave lectures in 11 provinces; his itinerary included Shenyang, Tianjin, Jinan, Taoyuan, Guangzhou, Hubei and Hunan. He gave nearly 200 lectures, to audiences totalling tens of thousands. Chinese newspapers gave him wide coverage. During his visit, 140,000 copies of his lectures were sold; they were reprinted until the 1950s.

Hu was his main translator; if he had other commitments, a fellow student from Columbia, Wang Wei, stepped in. In accompanying Dewey, Hu had the opportunity to meet leaders in the provinces and cities where he lectured, including warlords. It greatly widened his personal network. One phrase that became popular during the visit was that China needed Mr De and Mr Sai – "Democracy" and "Science".

One reason Dewey stayed so long was his fascination at what he saw; it was a social and intellectual ferment he had never witnessed before. He described the intellectual landscape as vexed by "confusion, uncertainty, mutual criticism and hostility among the various tendencies." Young Chinese, he said, had "all kinds of contradictory ideologies".

Such a nationwide tour was unprecedented in China, before or since. Since the late Qing period, foreigners, especially westerners, had worked as teachers. The first president of BU, in 1898, was William Alexander Parsons Martin, an American Presbyterian missionary; he spoke and read Chinese fluently. But no foreigner had criss-crossed the country, spoken to thousands of people and had his works printed and sold. This was made possible by the openness of the government and the enthusiasm and organisational talent of Hu and fellow alumni of Columbia; they were inspired by Dewey.

His message to Chinese was the same as he delivered at home – transform the country through education and social reform, not revolution: allow freedom of thought, analyse problems scientifically and find solutions suitable to domestic conditions – in China, not the West. Political reform alone could not save China; it needed reform from the bottom up, especially an educational system that best utilised its human resources. His advice to China was "the third way", between radicalism and conservatism.

"There are men who propose grandiose schemes by means of which they would reconstruct the world once and for all," said Dewey. "But I, for one, simply do not believe that the world can be reconstructed totally and on a once-for-all basis; it can only be reconstructed gradually and by individual effort ... The society which we desire is one in which there is a maximum opportunity for free exchange and communication."

Historically, education in China had been used to impart rote learning. That should not be so, said Dewey; its role was to train students to think for themselves, develop their individualism and curiosity and participate in civil society. Dewey's message reached hundreds of thousands of individuals. But his influence on China's history turned out to be limited; to implement its ideas required a political and legal stability that China did not have. Another Westerner, Karl Marx, had far greater impact – he lived during the Qing dynasty, from 1818 to 1883, and never visited China.

Another important guest lecturer at BU was British philosopher Bertrand Russell, who arrived in China in October 1920 and stayed for nine months. He toured the country and gave 63 public lectures. In 1924, it was the turn of Rabindranath Tagore, the Indian author who won the Nobel Prize for literature, to visit BU. Many of his works had been translated into Chinese.

These visits by distinguished foreign scholars showed the desire and energy of Hu and his colleagues to bring the best foreign learning to the people of China. They also demonstrated the openness of the Beiyang government. The governments before 1911 and after 1949 would not have permitted such nationwide visits.

A Doll's House and the Status of Women

In the summer of 1919, student societies across China staged a new play *The Greatest Event in Life*; it quickly became a hit. The heroine was a 23-year-old lady, Tian Ya-mei, who refused to accept the marriage arranged by her parents. The play ends with Miss Tian leaving a letter to them and escaping to join her boyfriend waiting in his car outside. The author of the play was Hu Shih; he had published the script in the March issue that year of *New Youth*. It was inspired by and based on *A Doll's House* by Norwegian playwright Henrik Ibsen, a play which Hu had read in 1914.

The heroine of that play, which premiered in Copenhagen in December 1879, was Nora Helmer, wife of a bank manager. It described her awakening to the fact that her life had been controlled by her father and her husband. She came to realise that she did not love her husband; the play ended with her slamming the door of the family home and leaving him. It was one of the first plays in the world to deal with the emancipation of women.

This was one of many issues which Hu had adopted during his seven years in the U.S. He had met independent women like Miss Williams and her friends; they were well educated and self-confident and regarded themselves as the equal of men. While society and their parents wanted them to marry, it was, for them, an option, not an obligation; they had a choice, if not complete freedom, in selecting their marriage partner.

Hu also saw in the U.S. the great contributions women made to society in the fields open to them – such as doctors, lawyers, teachers, nurses, midwives, writers and dress-makers. Miss Williams herself was a good example – first an avant-garde artist and then a librarian at an Ivy League university. Alice, wife of John, Dewey, was another example of a sophisticated woman fully engaged in society; her field was education.

Hu compared the women of China – deprived of an education and excluded from the professional workforce, with millions having their feet bound as a sign of servitude: what a waste of this rich human resource.

In June 1918, *New Youth* devoted an entire issue to translations of Ibsen's work. Hu introduced the issue with an article called "Ibsenism". He praised it for describing social issues on the stage and promoting individualism. In *The Greatest Event in Life,* Miss Tian meets her boyfriend, Mr Chen, while she is studying overseas, and falls in love with him. Her parents strongly oppose a marriage. A fortune teller informs her mother that the birthdays of the two do not match; and the two surnames do not fit the ancestry of the Tian family.

Miss Tian tries hard to convince them to change their mind, but in vain. At the end, she leaves them and joins her beloved. His play and *A Doll's House* created a sensation among young people. It persuaded many to follow the example of Miss Tian and leave the security of their families and take a partner of their own.

Another women's issue addressed by Hu was the hypocrisy in China, as in many countries, over the sexual freedom accorded to men but denied to women. Wealthy men in China had concubines in addition to their "official wife"; they could visit prostitutes and remarry if their wife died. Society condoned this behaviour. A woman, on the other hand, was supposed to

Hu Shih (second left) with Indian scholar Rabindranath Tagore (centre), whom he invited to China, in 1924.

be sexually faithful to her husband and no-one else, even after his death.

In an arranged union, women often did not see their husband before the marriage, as in the case of Hu and Miss Jiang. In the July 1918 issue of *New Youth*, Hu wrote an article "The Question of Chastity". He said that a man who was not faithful to his wife could not demand that she stay faithful to him. Society permitted concubines and prostitutes and honoured women who went to great lengths to remain chaste.

The article attacked the practice of a young woman giving up the right to marriage if her fiancé died before the wedding or her husband died early. In some cases, the widow committed suicide after losing her husband. Hu said that men should adopt the same attitude to women as they expected women to adopt to them. Chastity is equally binding on both sexes.

"Chastity reflects the manner in which husband and wife treat each other," he wrote. Whether a widow should remarry was entirely a matter for her to decide; the rule that she should remain a widow was unreasonable.

In 1922, Hu invited Margaret Sanger, an American feminist and founder of the modern birth control movement, to speak at BU. Like Dewey, she was lecturing in Japan. Born in September 1879, Sanger was the sixth of 18 children conceived by her mother over a period of 22 years; of these, 11 survived. Her mother died aged 49.

Trained as a nurse, Sanger opened the first birth control clinic in the United States in October 1916. She believed that, to be equal with men and have a healthy life, women should determine whether and when to have children. Shortly after the clinic opened, she was arrested and found guilty of breaking a law that banned distribution of contraceptives. She was sentenced to 30 days in a workhouse.

In 1921, she founded the American Birth Control League. At BU, about 2,000 attended Sanger's lecture on "Birth Control", with Hu as translator. Thanks to him, the students were able to hear a woman who was a global pioneer in her field; her ideas were controversial in the United States and Europe as well as in China.

Two Chinese women's magazines published Hu's translation of Sanger's lecture in its entirety. Chinese working in birth control adopted Sanger's theories and methods. A century later, the women of China have education, rights and freedoms which Hu's mother and wife could never have imagined. Hu and his colleagues made an important contribution toward this progress.

"I did not want a son, he came by himself"

In the summer of 1918, Miss Jiang moved to Beijing to live with her husband. Thanks to his high salary, they had a spacious home with servants, as well as a stream of relatives visiting from Anhui.

A Manchu lady managed the house, helped by a maid and a cook. Miss Jiang enjoyed mah-jong; Hu joined her and the other players from time to time. Mostly, he worked in his study. Their first son, Zu-wang was born in 1919. Zu-wang means "the hope of grandmother", in memory of Hu's mother who had died a year before. Daughter Su-fei was born in 1920. She was named after Sophia Chen, a close friend of Hu and the first woman professor at a Chinese university. The two met at Cornell University in 1916. Chen graduated from Vassar College in Poughkeepsie, New York with a BA in history and returned in 1920 to China; she taught Western history at BU.

In 1925, Su-fei died of tuberculosis, leaving her parents broken-hearted. Their second son, Si-du, was born in 1921; si-du means "remembering (John) Dewey".

Miss Jiang controlled the finances. The family spent little – the main expenses were Hu's books and imported cigarettes and his wife's losses at mah-jong. He was well paid, especially for someone so young – but they had many relatives who needed money.

After Zu-wang was born, Hu wrote him a poem, *My Son*. "Actually, I did not want a son. The son came by himself. We cannot hang the sign of 'no descendants'! If a flower blooms on a tree, the flower may accidentally fall. That fruit is you, that tree is me."

In the context of its time, this was a revolutionary poem. The first duty of every Chinese man was to marry an appropriate bride and present a son to his grandparents.

From his birth, the son took on the responsibility of caring for his parents and grandparents; the debt he owed them he could never repay. Young people Hu had met in the U.S, rejected this idea. They said that you did not need to marry; and, if you did, you could choose not to have children. In that way, you could break this never-ending cycle of duty and obligation.

Hu wrote articles exploring these concepts and rejecting the arranged marriage that was the norm for Chinese people. One of the most basic human rights should be the right to marry someone of your own choosing. But now Hu found himself following the tradition, marrying a woman he would certainly not have chosen. He was a creature of two eras, the past and the present, the Qing dynasty and Western liberalism.

In 1923, he took a year of sick leave from BU and spent seven months in the south, including five in Hangzhou, close to the West Lake, one of the loveliest places in China. During this period, he had a passionate affair with Tsao Cheng-ying, daughter of a concubine of an Anhui merchant and a cousin of Hu's by marriage. Through her own hard efforts, she had earned an education; she divorced her husband after he took a concubine. She was the kind of independent, educated woman Hu had known in the U.S; she was the opposite of his wife.

Tsao became pregnant with Hu's child. That winter he asked Miss Jiang for a divorce. She was angry and upset; the two quarrelled fiercely.

Jiang Dongxiu with their three children: from the right, Zuwang, Sidu and Sufei.

According to one account, she threatened to kill herself and her two sons with a kitchen knife if he divorced her. Hu backed down and agreed to stay in the marriage; he persuaded Miss Tsao to have an abortion.

Divorces were rare in China and not well regarded in society. If Hu had left his wife, it would have seriously damaged his reputation as a public figure. Later in life, he had affairs with other women – but never again asked his wife for a divorce.

Hu Shih and friends in Hangzhou in 1923. He is fourth from the left, and Tsao Cheng-ying second on the right.

Hu later helped Miss Tsao with her education. In the autumn of 1934, thanks to his recommendation, she was accepted by the agriculture faculty of Cornell University – the one he had dropped out of. He asked Miss Williams to look after her there. After her return to China with an M.A. in 1937, Tsao became the country's first lady professor of agriculture, at Anhui university.

Sources for Chapter Four

Website of Beijing University

Young Hu Shih 1891-1917, by Tang Yan, Spring Hill Publishing Company, Taipei city, first edition June 2020.

A Pragmatist and his Free Spirit, the Half-Century Romance of Hu Shih and Edith Clifford Williams, by Susan Chan Egan and Chou Chih-p'ing, Chinese University Press, Hong Kong, 2009.

If Not Me, Then Who? Hu Shih Volume One, Jiang Yong-zhen, Linking Publishing Company, Taiwan, first edition January 2011.

Hu Shih Oral Autobiography, as told to Tang De-gang, Yuanliu Publishing Company, Taiwan, published in November 2010.

"The Absence of Gender in May Fourth Narratives of Woman's Emancipation" by Yang Lianfen, Beijing Normal University, in *New Zealand Journal of Asian Studies 12, 1* (June 2010): 6-13.

Dewey and May Fourth China, State University of New York Press, Albany, 2007.

English Writings of Hu Shih, edited by Chou Chih-p'ing, Foreign Language Teaching and Research Press, Beijing, 2012.

From Pagan to Christian, Lin Yu-tang, World Publishing House, 1959.

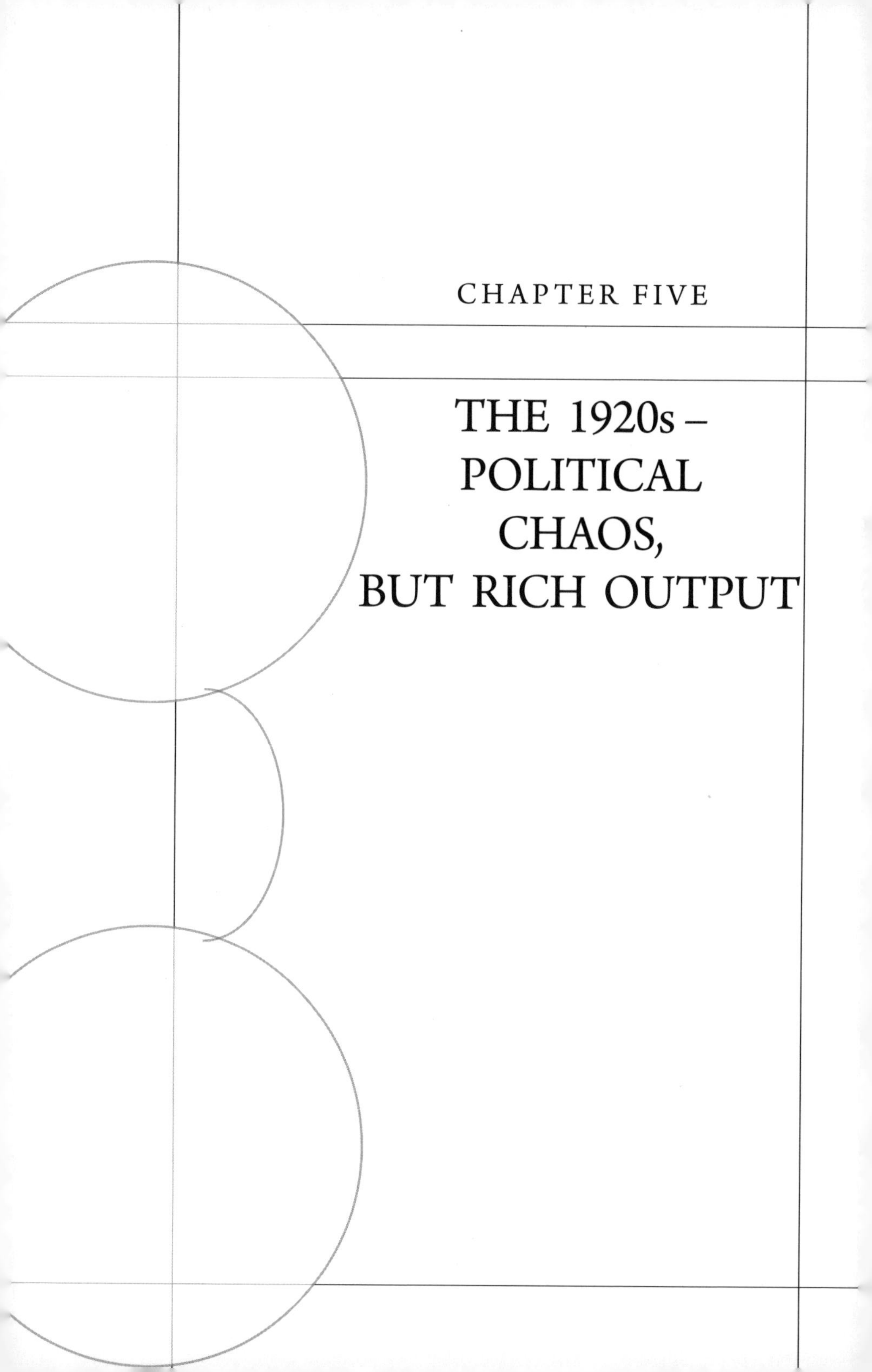

CHAPTER FIVE

THE 1920s – POLITICAL CHAOS, BUT RICH OUTPUT

For Hu, the 1920s were a period of intense intellectual activity in a period of instability and violence. For the first seven years, he lived in Beijing; during that period, the country had 19 Prime Ministers! Rival warlords fought each other for power, land and resources. This meant little or no money to pay salaries of professors at Beijing University. In 1923, its president Cai Yuan-pei resigned in protest and left China for three years.

In 1927, Hu judged it too dangerous to live in Beijing; with his family, he moved to Shanghai where they stayed for three years. He was dismayed by the outcome of the Xinhai Revolution and the May Fourth movement. They did not lead to the modern, democratic government he had hoped for.

The decade saw the rise of two political parties which both promised to end the instability and unify the nation – the Nationalist Party of Chiang Kai-shek and the Communist Party, founded in 1921. Two of its founders were Hu's former colleagues at BU.

In the midst of this chaos and instability, Hu continued to be immensely productive. He wrote books and articles on a wide range of subjects, including literature, philosophy and history as well as social and political issues. Major magazines and journals published his writings, giving him a national audience. His books included *An Outline of the History of Chinese Philosophy*, published in 1922, and *History of Vernacular Literature*, published in 1928.

At BU, he taught some of the best minds in China. Through his teaching and writings, he influenced thousands of Chinese, both intellectuals and the wider public. In 1926 and 1927, he spent nine months in Britain and the United States. He spoke at elite universities, met leading citizens and conducted research on Zen Buddhism, a subject that greatly interested him.

On March 14, 1920, Hu Shih visits a Buddhist temple in the Western Hills of Beijing with friends. From the left, Chiang Monlin, Cai Yuanpei, Hu Shih and Li Dazhao.

Seeking Funds

Beijing University could not be independent of the chaos surrounding it. From the start of 1921, the staff of BU and seven other colleges and universities in the capital received no salaries from the government. On March 15, the eight organised "A Joint Council of Staff Delegates"; it campaigned for their salaries and other funds for the universities. When professors and students took to the streets to protest on June 3 that year, they were beaten, shot and bayoneted by the guards of the president.

The campaign of the professors against the government went on for several years. In January 1923, Cai Yuan-pei resigned as president of BU, to protest the arrest of the Minister of Finance on trumped-up allegations of bribery; he went to live in France and did not return to China until 1926.

During this difficult period, Hu and his colleagues were greatly helped by subsidies from the China Foundation for the Promotion of Education and Culture. The CF was set up in 1924 to manage US$12.5 million from the United States government. This was a second tranche of money from the Boxer Indemnity, in addition to the one which paid scholarships to Chinese students like Hu to go to the U.S.

The Chinese government appointed a board of trustees with 10 Chinese and five Americans to manage the fund; its constitution said that it was self-perpetuating – a wise move, to save it from government interference. At that time, US$12.5 million was a substantial sum of money. Its money helped to rescue BU from financial collapse. It subsidised the payroll of its teachers and staff. It also established the National Library of Beijing and funded its building and purchase of rare books.

In 1923, with several friends, Hu set up the Crescent Society, a cultural group. The name came from a poem by Rabindranath Tagore. The main organiser was Xu Zhimo, a close friend of Hu and colleague at BU. Born into a rich family in Haining, Zhejiang on 15 January 1897, Xu studied in the United States and Britain; on his return, he went to teach at BU. He became a leading figure in China's modern poetry movement. The society put on plays, published magazines and acted as a platform for poets and writers in the vernacular and new forms of expression. Xu introduced western romantic forms and worked to free poetry from its traditional forms. When Tagore visited China, Xu was one of his interpreters.

As he had in the U.S., Hu maintained links with a wide range of people, Chinese and foreign. He was a friend of Reginald Fleming Johnston, English teacher of Pu Yi, the deposed emperor. A photograph of 1923 shows Johnston taking Hu on a tour of the Forbidden City. Photographs in 1924 and 1925 show Hu at BU meeting friends from Japan, including scholars and a member of its Parliament.

The activism of the Beijing students and the violent reaction of the authorities posed a dilemma for their professors. Should they support them and, if so, how much? Should they join the protest marches and risk injury, arrest and inability to continue teaching? Many BU professors had studied in Japan, the U.S. and Europe, in countries where such protests were permitted and considered a normal part of university life.

Hu himself was a moderate. While he often supported the demands of the students, he opposed the disruption of formal education – students should be in the classroom and the library, not on the streets. He discouraged activism that had no clear or achievable goals. He followed the philosophy of his mentor Professor Dewey that education was the most important way to create social progress and reform. To become a modern nation, Chinese needed to change their way of thinking.

"The greatest source of evil in China today is not warlords and bad officials, it is the lazy and superficial way of thinking," he wrote during this period. "It is the superstition of relying on heaven for your food and the attitude of indifference – watching the fire from the other side of the river. These things are our real enemies. Our evil politics is the result of over 2,000 years and the culture and thinking they have created. Everyone must work hard to destroy them."

Founding the Communist Party

Two of Hu's friends and colleagues at BU were founder members of China's Communist Party – Chen Duxiu and Li Dazhao. Because of the prominent role Chen had played in the May Fourth protests in Beijing, he was sentenced to three months in jail. On his release, he left the city and moved to the safety of the French Concession in Shanghai. There he continued to edit the *New Youth* magazine; it become increasingly leftwing, too much so for Hu.

In May 1920, Chen met two agents of the Communist International (Comintern) sent from the Soviet Union. Chen became secretary of a provisional central committee of the Communist Party. In December 1920, Li set up the party group in Beijing.

In the summer of 1921, 21 of the 53 party members were students or faculty of BU. On July 23 1921, the party held its first plenary meeting in the Huangpu district of the French Concession in Shanghai; the venue was a residential building, with 13 Chinese and two Comintern delegates present. Although unable to attend, Chen was chosen as secretary-general. Li Dazhao was also absent but is considered one of the founders.

Chen had much in common with Hu. Both came from a well-established family, received a traditional education and then studied abroad. As editor of *New Youth* magazine, Chen published the article by Hu on promoting the vernacular that made him well-known in China. Chen became dean at BU and Li Dazhao its head librarian, at the same time Hu was teaching there. The three men were outstanding intellectuals; all advocated sweeping reforms for China.

But Chen and Li did not have Hu's long and positive experience in the

United States, nor a professor as charismatic as John Dewey. The two did not enter the mainstream society in Japan as Hu had in America. In Japan, they devoted most of their time and energy to organising protests against the Qing government – in Li's case, so much so that Waseda University in Tokyo expelled him for his long absences from class. While Hu saw much in the U.S. for China to follow, Chen and Li did not see Japan as a model for China. In personality too, they were different – Hu was deliberate and analytical, Chen volatile and emotional.

The role of the Comintern agents sent from Moscow – Grigori Voitinsky, Yang Minzhi and Henk Sneevliet (pseudonym Maring) – was critical in setting up China's Communist Party. They brought the powerful credential of a party that had overthrown an imperial dynasty and set up a form of government never seen before in the world. They brought with them the ideology, organisation and systems needed to set up a nationwide political party.

In 1925, they sent to Moscow the first 340 Chinese, mostly members of the Communist Party, to study at the Sun Yat-sen University set up to train revolutionaries. But neither Li nor Chen survived long in the revolutionary inferno. On March 18, 1926, Li organised an anti-government demonstration in Beijing; government soldiers fired into the crowd, killing 47 and wounding over 200. On the most wanted list, Li escaped and took refuge in the Soviet embassy in Beijing.

In April 1927, Zhang Zuolin, a pro-government warlord from Manchuria, ordered his soldiers to enter the embassy and arrest Li and his associates sheltering there. On April 28, the soldiers hanged Li and 19 of those held with him.

The embassy remains today in the northeast of the city, next to the first ring road; one of the largest diplomatic sites in the world, it has apartment buildings, a school, sports facilities, spacious grounds and a bomb shelter – an excellent place to take refuge.

For his part, Chen was in 1927 dismissed as head of the CCP, after he fell out with the Comintern and Mao Tsetung. The party expelled him in 1929. The Nationalist government held him in prison from 1932 until the outbreak of the Sino-Japanese war in 1937. In 1942, he died in Jiangjin, a town west of Chongqing, aged 62. If Chen and not Mao had remained head of the Communist Party, the history of China would have been very different.

While Hu did not share the political convictions of his two former colleagues, he remained friends with them. In 1919 and 1922, he lobbied with the government to obtain Chen's release from prison. In 1934, Hu visited Chen in prison. After his death, he wrote an introduction to a collection of his late writings. In 1930, he dedicated the third volume of his *Collected Essays* published that year to Li Dazhao. Hu himself tried to stay out of politics. He was happy to express his opinions but did not want to join a political party. He saw his mission as reforming China through education, literature, philosophy and culture. Such neutrality became increasingly difficult in the bitter and deadly political wars of the 1920s.

Two parties became most powerful – the Nationalists of Chiang Kai-shek and the Communists led by Mao after the purge of Chen Duxiu. Chiang launched a successful military campaign against the warlords who controlled much of China. On June 8, 1928, his forces captured Beijing; Zhang Zuolin, then President of the government in Beijing, acknowledged Chiang's authority.

Chiang established his new administration in Nanjing; he chose it as the new capital, to replace Beijing. On October 10 1928, he formally proclaimed a new National Government at Nanjing. Its authority stretched from Guangzhou in the south to Shenyang in the northeast.

"I criticised Anarchism, Socialism and Bolshevism"

From the early 1920s, Hu opposed Communism, a stance he maintained through his life. He considered that the party's slogans – like those of feudalism, capitalism and imperialism – misled people as to how to solve China's many and complex problems; its sloganeering was illogical and authoritarian and the antithesis of the scientific method and analytical thinking he advocated.

"I criticised those ideologies that were blindly accepted, such as Anarchism, Socialism and Bolshevism." Hu said it was very easy to talk about these "isms" and that they were used by ambitious people for their own personal benefit. China did not need the revolution of violent despotism nor a revolution that fought violence with violence, he said. What it needed was detailed analysis of individual problems and to solve them through trial and error. He summarised his thinking with a phrase he often used: "be bold in suppositions, be careful in seeking evidence".

He believed that the greatest danger of the "isms" was that they satisfied people and made them believe they had found a total solution. But many did not want to listen to Hu's analysis. They wanted quick and radical solutions to poverty, illiteracy, disease, corruption and foreign interference in China.

"In fact, in 1919, Chen (Duxiu) still did not believe in Marxism," Hu wrote. "In his early writings, he plainly opposed Socialism ... In 1918 and 1919, Li Dazhao had already written articles praising the Bolshevik

Revolution in Russia. By comparison, Chen was a late convert to Socialism. After unfortunate experiences with the Beijing police, Chen left the city. From January 1920, he left our circle at BU and his old friends at the *New Youth* magazine and became increasingly distant. Under his editorship in Shanghai, it turned into a magazine promoting the workers' movement and later promoting Communism. Finally, in the French concession of Shanghai, it was closed down."

Hu had different ambitions to those of his two former colleagues. In December 1919, he set out the objectives of his Literary Revival in an article entitled "The Significance of the New Wave of Thinking". They were: to research problems, especially the pressing problems of the day; import from abroad theories suitable to what was being researched; arrange China's national heritage, including 3,000 years of classical writing that was broken and disorganised, and use scientific methods for systematic arrangement; these would remake China's civilization.

These were ambitious targets but ones which people could achieve, he believed, without diving into the snake pit of politics.

He maintained his opposition to Communism for the next 30 years. In December 1948, the People's Liberation Army was at the gates of Beijing. Many of his fellow intellectuals, who detested the Nationalist Party, chose to await the arrival of the new rulers. Chiang Kai-shek sent a plane to Beijing to rescue those scholars who wanted to leave. Hu knew well what his future would be in the People's Republic. So, he took the plane – and never returned to the mainland.

After he took power in 1949, Mao Tse-tung took his revenge against Hu for his opposition to the Communist Party and his refusal to stay in Beijing. He launched a nationwide campaign against him; for four years during the 1950s, newspapers and magazines printed millions of characters to denounce him. His second son, Hu Si-du, who had remained in Beijing in 1948, publicly denounced his father. In the newspapers, he described his father as "an enemy of the people", "my enemy" and "a running dog of imperialism".

In 1957, during the Anti-Rightist movement, Si-du was designated a rightist. On September 21, 1957, he hanged himself; he was aged only 35.

"Collection of the Zen Monks"

In 1922, the British government decided to follow the good example of the United States and set up its own Boxer Indemnity Fund; it was U.S. money from this fund that had allowed Hu and hundreds of other Chinese to study there.

The British government set up a committee to advise how to spend the money. It had 11 members, eight prominent British citizens and three Chinese, including Hu Shih and Ting Wen-chiang. A graduate of the University of Glasgow, Ting was one of China's first professional geologists. Hu and Ting accompanied members of the committee to Tianjin, Shanghai, Nanjing, Hangzhou and other cities while they considered how and where to spend the money.

To complete his work with the committee, Hu applied for and received a year's leave from BU. In July 1926, he set off by train from Beijing and, for the first and only time in his life, visited Moscow, birthplace and centre of the Communist revolution. He stayed there for three days. In a letter to his friend Xu Zhimo in August, this is what he wrote about the visit: "I felt the people there have a seriousness of purpose, the atmosphere is one of earnestness and determination. On a visit to the Museum of the

Revolution, exhibits from the revolutionary movement there between 1890 and 1917 made me feel that we ought to be shamed to death ... we have fallen far short of the responsibilities entrusted to us by our positions."

In September, he wrote another letter to Xu about the Soviet Union. "The people surrounding Lenin are educated and experienced, but even they did not descend from heaven. Moreover, there is only subservience or defiance under a dictatorship, there is no right or wrong – the piglets (helping themselves to the public trough) today (not limited to senators) will be the same kind of people who thrive under a dictatorship. Under such a system, there is no room for people like us who insist on independent thinking. If we want to save our country, we need to improve the education and mind-set of the people, there is no shortcut." This was Hu's mindset in December 1948 when Chiang sent the aeroplane to Beijing to rescue him.

After the members of the committee reached London in August, they held two formal meetings and completed a report; they submitted it to the British Foreign Secretary on 26 October, 1926. In the end, the government spent the money to purchase British-made bridges, locomotives, rolling stock, rails and other material for China's railways; in addition, it gave £465,000 to Hong Kong University and the Universities' China Committee to educate Chinese students and promote closer cultural relations between Britain and China.

For Hu, the visit to Europe was also an excellent opportunity to research Chinese Buddhism, a subject that interested him greatly. For this, he had to go to the Bibliotheque Nationale in Paris and the British Museum in London. Both had priceless manuscripts from caves in Dunhuang in Gansu province in west China.

In his oral autobiography, Hu recounts the astonishing journey of these manuscripts. In 1900, a Taoist priest in Dunhuang was sweeping the earth when he discovered a large stone door next to wall paintings. He opened the door and found inside a trove of 10,000 scrolls. Since the priest was illiterate, he could not read the scrolls. He sold a few of them to local villagers in exchange for medicine to cure his toothache. Since the population there was small, he sold only a few. In 1907, a Hungarian-born British archaeologist, Sir Aurel Stein, was visiting the area and heard of the discovery. When he saw the scrolls, he realised that they were Buddhist writings from the fifth to the 11th century A.D. He paid the priests 70 silver dollars and filled 7 trucks with the scrolls. First, he took them to India and later gave them to the British Museum.

In 1908, Paul Pelliot, a French Sinologist who had heard of Stein's discovery, also went to Dunhuang. He could read Chinese and some Central Asian languages. After choosing carefully the scrolls he wanted, about 3,000-4,000, he took them to Beijing, where he invited Chinese scholars to help him understand them; then he donated them to the Bibliotheque Nationale in Paris.

The scholars concluded that monks had hidden the scrolls inside the caves during an emergency, probably a war, and sealed the doors before they escaped; the arid climate of Dunhuang had miraculously preserved them over many centuries.

In Paris, Hu read 50 Buddhist scrolls of the Tang dynasty and, in London, 100. They gave him important material on the history of Zen Buddhism, especially of the eighth century. Hu made copies of the texts. In 1930, he published in Beijing a book of the materials, entitled *Collection of the Zen Monks.*

The British Museum collection contains over 45,000 manuscripts and printed documents on paper, wood and other materials from different sites along the Silk Road. Of these, 20,000 are in Chinese, with the others in Tibetan, Sanskrit, Uighur, Mongolian and other languages.

The first translation of Buddhist scriptures into Chinese came in the second century A.D. Since then, Buddhism has been Sinicised and become part of the national culture; it is not regarded as a "foreign" religion. The development of Buddhism in China remained an important subject of research for the rest of Hu's life.

In Britain, Hu was a celebrity, giving lectures at leading universities and learned societies. Many members of the British elite invited him to meetings, including: Sir Charles Addis, chairman of the Hong Kong and Shanghai Bank; Randall Davidson, the Archbishop of Canterbury; Ramsay MacDonald, leader of the Labour Party and a former Prime Minister; and historian Arnold Toynbee. Such a warm welcome showed two things – the widespread interest in China and its future and the fact that, more than any other Chinese, Hu could interpret the country for them. He had the charm, English fluency, intellectual weight and experience of public speaking that made him the ideal dinner guest and speaker.

At the end of 1926, he moved on to the United States, where he was equally in demand. He gave lectures on Chinese philosophy at Columbia and Harvard Universities and took part in a Foreign Policy Association debate at the Astor Hotel in New York on the "Future of China"; 1,500 people were present in the room, with more than 50,000 listening to a live broadcast. While in New York he spent time with Professor Dewey and his family and he submitted his thesis to Columbia University, receiving his Ph.D.

In March 1927, he spent three weeks in Ithaca, the hometown of Cornell University and Miss Williams. It was their first meeting since he had left for China 10 years before.

"Japan has impressed me tremendously"

After over three months in the United States, on April 12 he sailed from Seattle for Japan. During a stay of 23 days; he met Japanese scholars of China and Buddhism. He visited Kyoto, Nara, Osaka and Hakone.

In a letter to Miss Williams on May 17 while he was still there, he wrote: "Japan has impressed me tremendously. Such a great progress has been made during these 10 years! In Tokyo and some other modern cities, the rickshaw is disappearing. And that is without the benefit of Buddhism or Confucianism or Christianity – but only as the natural result of material progress! What a lesson!"

He saw Japan as the Asian country that had most successfully adopted Westernisation. That first became clear in 1894-95, when China suffered a humiliating defeat at the hands of its small neighbour. Its success in modernisation, and China's failure, was a subject Hu often discussed in its writing and speeches. This is part of an English article he wrote in a magazine published in New York, in 1928. "Japan has achieved a modern civilization within a short time by an unreserved acceptance of the tools and machines of Western civilization. When Commodore Matthew C. Perry of the U.S. Navy knocked at the gate of Japan (in 1853), she was deep in her medieval slumbers ... In the face of imminent dangers of national humiliation and ruin, she did not trouble about her medieval religious and feudal morals, but went wholeheartedly into the work of equipping itself with all the new weapons of war, vehicles of commerce, machines of production and methods of organisation. In the course of half a century, Japan has not only become one of the greatest powers of the world, but has also solved a number of important problems which neither Buddhistic religion nor Chinese philosophy had been able to solve ... Japan has today 90 institutions of scientific and technological research and 30,000 engineers enrolled in the membership of her national engineering societies."

He returned to this theme in the first presentation to the Haskell Lectures at the University of Chicago in the summer of 1933, comparing the responses of Japan and China: "China has wasted fully a century in futile resistance, prolonged hesitation, spasmodic but incoherent attempts of reform and disastrous wars of revolution and internal strife. Today she is still displaying to the world the most pathetic spectacle of a once great national helplessly struggling to stand on its own feet again and groping desperately to find ways and means for the solution of her numerous and pressing problems created and complicated by the impact of the irresistible civilisation of the West."

He defined three factors present in Japan and absent in China that enabled that country to succeed – a powerful ruling class that produced the leaders of reform and modernization: the fact that this class was a specially privileged and highly trained military caste: the political development of Japan over 1,000 years that created a solid centre of gravity for change. But, during this time, China was controlled by absolute rulers who abolished projects of reform and modernisation, the most recent being the 100 Days Reform in 1898; they were cancelled by Empress Dowager Cixi Taihou.

The nine months away showed the breadth of Hu's global network of contacts, which few, if any Chinese, could match. In late May 1927, he returned to China and met again his wife and two sons. Miss Jiang played

Hu Shih (far left) visits a temple in Kyoto with Japanese scholar friends

no part in his intellectual and professional life, a situation both accepted.

Crescent Bookshop and Human Rights

After his return to China, Hu decided that it was too dangerous to live in Beijing. Many of his friends had already left the city. Because he had praised the Russian revolution, some accused him of being a Communist.

In the autumn of 1927, he moved with his family to the safety of the International Concession in Shanghai. He rented a three-storey Western-style house in Jessfield Road, near the home of Cai Yuan-pei. He started teaching philosophy at the private Kwanghwa University.

In April 1928, he became principal of China National Institute, the secondary school where he had studied 22 years before. He earned a comfortable living from university lectures and royalties from his collected works. During his three years in Shanghai, he continued his intense work schedule – seven chapters of the second volume of his intellectual history of China, articles on Buddhism and scholarly prefaces to novels. He also started a series of essays on his family and early life; they were later collected in a book *Autobiography at Forty*. In June 1927, he became a trustee of the China Foundation for the Promotion of Education and Culture (CF), an institution that would play an important role in the rest of his life.

Among those who also moved from Beijing to Shanghai was Xu Zhimo and other members of the Crescent Society. In 1927, they opened the Crescent Bookshop in Huanlong Road; Hu was chairman of the bookshop. In March 1928, they set up the *Crescent* literary magazine to publish poems and articles by leading authors, including Ba Jin, Ding Ling and Yu Da-fu. It mainly covered literary and cultural topics, with some articles on human rights. It published 43 issues before it closed in June 1933. This followed the death on November 19, 1931 of Xu Zhimo in an aeroplane crash in Taian, Shandong province on his way from Nanjing to Beijing; he was just 34. Hu wrote a poem to mourn him.

On December 1, 1927, Hu attended China's most important wedding of the year – Chiang Kai-shek and Song Mei-ling. They held a private Christian wedding at the Song family residence in Seymour Road in Shanghai in the morning. In the afternoon, they went to the Majestic Hotel for a formal ceremony attended by more than 1,000 guests, including senior KMT ministers and celebrities. The invitation to Hu was a sign of his status in the official world; he was a scholar with a national reputation. Chiang was the most powerful man in China; the two men did not talk during the reception. Hu would soon see more of him.

In 1930, Hu wrote an open letter criticising the Nationalist party for a proposed law that would give it the power to decide who was "counter-revolutionary" – a crime which carried the death penalty. He gathered his articles and those of his friends on human rights into *A Collection of Essays on Human Rights,* published by Crescent Society in 1930. These criticisms resulted in threats and menaces. This toxic atmosphere persuaded him to resign as president of the China National Institute in May 1930; it was the only way to secure official recognition for the school. In an article that year "Introducing My Own Opinions", he expressed his despair 13 years after his return to China: "We must recognise that we are behind others in everything – in materials and machinery, in political system, in morality, in knowledge, in literature, in music, in art and in physique."

Translating Shakespeare

In July 1930, the CF set up a 13-member Committee on Editing and Translation, with Hu as chairman. It had two divisions – history/literature and natural science. This was something for which Hu was ideally suited – translating major Western literary and scientific works into Chinese; it was a project that seized his imagination. It was an excellent reason to remain in China.

One remarkable result was the work of Professor Liang Shih-chiu. In 1931, he was head of the English Language Department of National Qingdao University. That year he accepted the offer to translate the works of William Shakespeare. It took him 37 years to complete the task – in 1968, when he was living in Taipei. He wrote 40 volumes and three million characters.

"The translation of Shakespeare was first suggested by Dr Hu Shih, chairman of the Translation Committee," wrote Professor Liang in a 1963 article, which was reproduced in English translation in *Renditions* magazine in 1974. "His plans were ambitious and the translation of Shakespeare was but one of them."

Other books translated were *Novum Organum* by Francis Bacon and *Discours de la Methode* by Rene Descartes, on philosophy, as well as literary works by Alexandre Dumas, Joseph Conrad and Thomas Hardy.

"The work of this Translation Committee gradually expanded until it came to an abrupt end at the outbreak of the Sino-Japanese War," said Professor Liang in the 1963 article. In 1936, the Commercial Press published the first seven of his translations, including *Hamlet, Macbeth, King Lear* and *Othello.*

His work was disrupted first by the war with Japan and then the Chinese civil war and runaway inflation. He escaped to Taiwan with none of his reference materials and had to build his library again from scratch. "I want to express my admiration for and fond memories of Dr Hu, without whose enthusiastic initiation I would never have embarked on the translation of Shakespeare at all. Dr Hu is not a Shakespearean scholar, but he recognised the importance of translating Shakespeare and took it upon himself to plan the whole project responsibly and carefully."

These translations, of many scientific and literary works, are part of Hu's rich legacy to China.

Return to Beijing University

China was increasingly dangerous, with bitter political infighting and the constant threat of attack by Japan. Hu received many invitations to teach at leading universities in the United States, with salaries up to 10 times what he had earned at BU. It was tempting to leave China for the peace and tranquility across the Pacific Ocean. At home, his favourite place was Beijing University. In 1930, he received an invitation to return there from old friend Chiang Monlin, a classmate at Columbia University.

In 1908, Chang had gone to the United States in 1908 and stayed for nine years. Like Hu, he obtained a Ph.D at Columbia under the guidance of Professor John Dewey. From 1919 to 1927, he was acting president of Beijing University and, from 1928 to 1930, Minister of Education. Then he returned as president to BU. He invited Hu to serve as Dean of the Arts Faculty. With the formation of a national government in Nanjing, stability had returned to Beijing. It was safe for Hu and his family to live there. On November 26, 1930 they left Shanghai by train. They went to a courtyard home at 4 Xiamiliang ku which he had rented. On December 17, his friends held a party there to welcome him home.

In a letter to Miss Williams on March 25, 1931, Hu explained why he did not go to the U.S., as she strongly wished he would: "I felt in the first place, like a deserter to leave China at the present moment when I am popular abroad and most unpopular in my own country. Secondly, I started to write the second and third volumes of my *History of (Chinese) Philosophy* last March … I have moved back to Beijing to live, and for the first time in six years, the two parts of my private library are coming together."

Another reason to stay was the foundation in June 1928 of Academia

Sinica as the research academy of the government, by Cai Yuan-pei. It was based on similar government institutions in Europe that pursue scholarship without teaching. Its objective was to conduct academic research, and publish and promote research. It set up Institutes of Engineering, Chemistry and Physics. Cai Yuan-pei was chairman of the CF as well as of Academia Sinica. This was the kind of institution Hu admired and wanted to be part of. From 1958 until his death in 1962, Hu would also be chairman of Academia Sinica.

Sources for Chapter Five

Beijing University website

Papers regarding the disposal of the British Share of the China Indemnity of 1901, September 19 - November 14, 1930 His Majesty's Stationery Office.

Website of China Foundation for the Promotion of Education and Culture

A Pragmatist and his Free Spirit, the Half-Century Romance of Hu Shih and Edith Clifford Williams, by Susan Chan Egan and Chou Chih-p'ing, Chinese University Press, Hong Kong, 2009.

If Not Me, Then Who? Hu Shih Volume Two, Jiang Yong-zhen, Linking Publishing Company, Taiwan, first edition January 2011.

Hu Shih Oral Autobiography, as told to Tang De-gang, Yuanliu Publishing Company, Taiwan, published in November 2010.

"On Translating Shakespeare" by Liang Shih-ch'iu, *Renditions magazine*, Autumn 1974.

English Writings of Hu Shih, edited by Chou Chih-p'ing, Foreign Language Teaching and Research Press, Beijing, 2012.

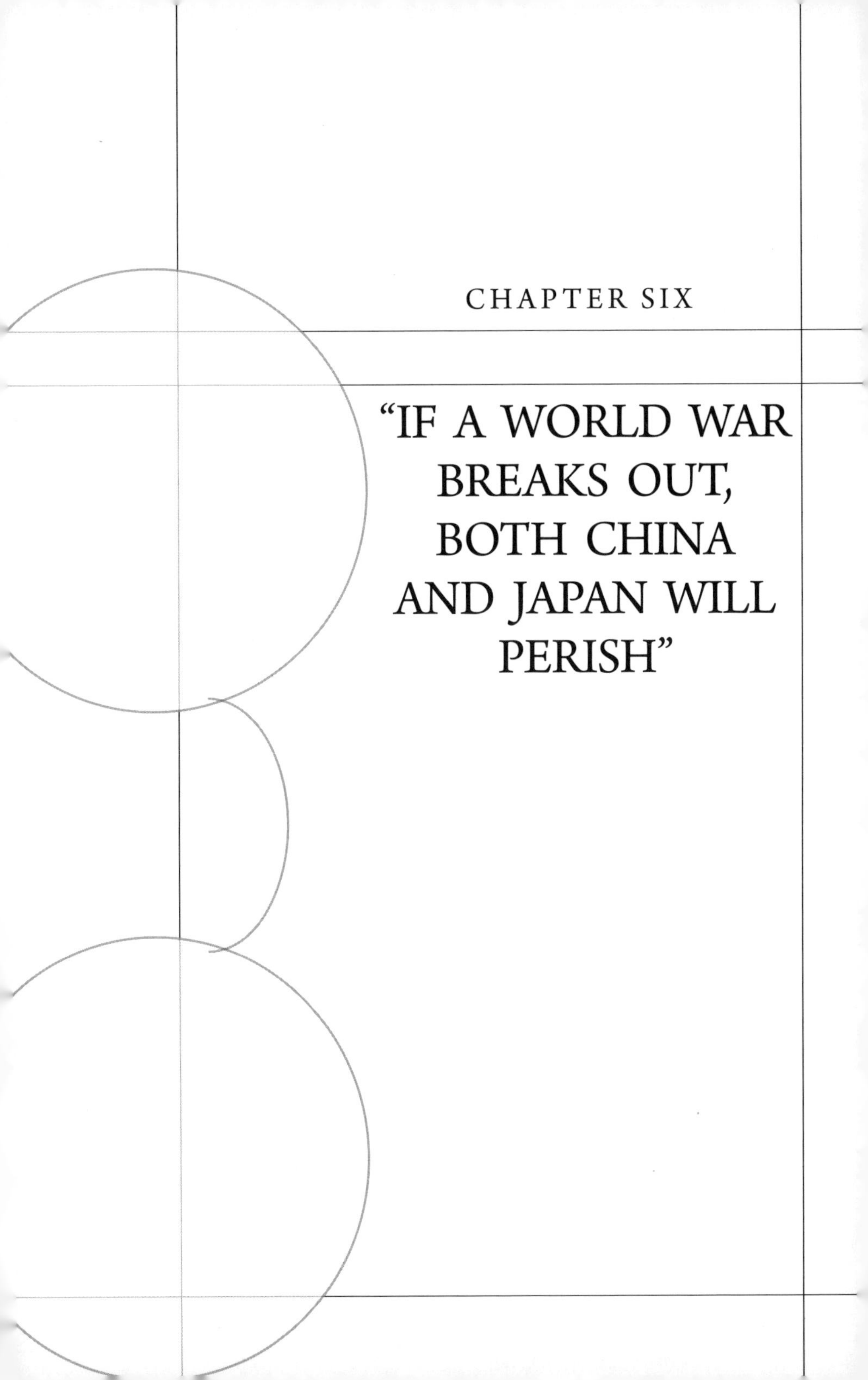

CHAPTER SIX

"IF A WORLD WAR BREAKS OUT, BOTH CHINA AND JAPAN WILL PERISH"

On the morning of September 19, 1931, the Japanese Kwantung Army shelled the Chinese garrison in Shenyang, capital of Liaoning province. The 500 Japanese troops, better trained and equipped than the 7,000 Chinese of the garrison, needed less than 24 hours to take over the city, with the loss of two Japanese and 500 Chinese lives. It was the start of the occupation of three provinces of the Northeast, known as Manchuria. The Kwantung Army completed the operation in five months, although its 50,000 soldiers were outnumbered three-to-one by the Chinese troops in the region.

In Nanjing, 1,600 kilometres away, the national government, established only three years before, watched helplessly. It did not have the equipment, firepower and aeroplanes to defeat the Japanese attack.

Like everyone else, Hu heard the news from Manchuria with alarm. Better than most Chinese, he understood the military and industrial strength of Japan and how much weaker was his own country. He also knew that this was a national crisis such as China had never faced. The Western imperial powers wanted economic benefits and preferential treatment for their firms and citizens in China. But they did not want to take over the country, as some Japanese generals did.

This changed Hu's conviction that he should stay out of the government and politics. In 1931, he accepted an invitation to join the National Economic Commission as an adviser. This was an institution set up in April 1931 to take overall charge of the economy during the Great Depression. The invitation to join came from President Chiang Kai-shek himself and was relayed by Song Zi-wen, Minister of Finance and Chiang's brother-in-law. Like Hu, he was a graduate of Columbia University.

Hu had often criticised the Nationalist government. But he decided that, in this national crisis, having a strong, if autocratic, central government was the best option. In November 1932 in Wuhan, after giving a lecture at the city's university, he had dinner with Chiang and Song Mei-ling; it was the first time he had met them in private. He told Chiang of his concerns over political stability and asked about funding for China's education.

On March 13, 1933, Hu met Chiang again, this time in Baoding. Hebei province. The Kwantung army was occupying the province of Rehe to the north.

In his diary that evening, Hu wrote of Chiang's surprise at how quickly the Japanese had captured Rehe and the good quality of their intelligence. "I [Chiang] had not expected that they knew much more about the armies of Tang Yu-lin and Zhang Xue-liang than we did".

Asked if China could resist a Japanese invasion, Chiang said that it would need three months of preparation. After the three months, could China attack then? "In terms of modern warfare, we could not. We could only in certain places die bravely in defending our positions. Not allow one soldier to stay and then retreat. In this way, we would show the people of the world that we are not afraid to die."

Chiang said that Japan had refused his offer of talks in exchange for giving back Manchuria; they insisted on keeping their new state of "Manchukuo."

A comparison of the two militaries showed the accuracy of Chiang's judgement. The Imperial Japanese Army (IJA) had 4.48 million men, with 2,700 planes and a naval fleet of 1.9 million tons. China had 1.7 million soldiers, 300 planes and a navy with 46,500 tons.

Better than anyone, Chiang knew the quality of the enemy. In 1907, he had studied in a preparatory school for the IJA in Tokyo. After graduation, he served in the IJA from 1910 to 1911. He admired the discipline, physical training and obedience of the Japanese soldier and how the training instilled a sense of will and intensity. He also admired their custom of bathing in cold water and eating cold food. For the rest of his life, he followed habits he had acquired in Japan – a strict routine, rising at 05:00 and eating simple meals.

It was out of this sense of military weakness that, on May 31 1933, the Chinese government signed the Tanggu Treaty with the Japanese to create a demilitarised zone 100 kilometres south of the Great Wall from Beijing to Tianjin; troops from the two countries were moved out of the zone. This enabled Chiang to avoid a direct conflict and build up his forces.

During this "truce", Japanese war planes, sometimes up to a dozen in formation, flew over Beijing. They reminded the residents, including Hu in his study at home, of the threat on their doorstep. These meetings showed the high esteem in which Chiang regarded Hu and the fact that he saw a role for him during the coming conflict. Chiang's preferred option was first to defeat the Communist rebellion, which his military was capable of, and then face the Japanese. He called this strategy "Resist foreign aggression by first making peace at home".

Hu was greatly influenced by the Baoding meeting with Chiang. Here was the head of China's armed forces telling him that the country was ill equipped to fight the Japanese. So, what were the alternatives? One, he believed, was to negotiate with Japan to win more time for China to upgrade its military. Another was to persuade Britain and the U.S. to threaten war with Japan. Only they had the military power to defeat the enemy.

In a letter in August 1938 to Miss Williams, he said that, in 1935-36, I had offered to serve as China's Ambassador to Japan. "But, having been an 'independent' so long, my serious words were taken as a mere joking gesture. Perhaps there were people who did not want me to do it," he wrote. If he had been given the post, it would have been the best opportunity for him to negotiate a peace treaty between the two countries.

In a letter on June 27, 1935 to his friend Wang Shi-jie, Hu accurately predicted the next 10 years. "In the not-too-distant future, there will probably be a great war in the Pacific. We perhaps can be victorious. But we must be prepared for three-four years of terrible war. We must bite with the roots of our teeth and be ready that other countries will not enter the war for three years. We can only hope that, while we smash the enemy to pieces until he has nothing left, then the international community will enter the war and help us."

That is exactly what happened. China would fight Japan on its own for four-and-a-half years; it would take devastating military and civilian losses before the intervention of the United States in December 1941.

It was to bring about this intervention that Chiang sent Hu to the U.S. as ambassador in the autumn of 1938. In 1935, Chiang sent a statement to the Japanese government, through China's ambassador in Tokyo: "The war between China and Japan will be bound to be a world war and, if a world war breaks out, both China and Japan will perish."

He said that the two countries would benefit if they stuck together. Standing apart, each would suffer. His prediction turned out to be completely accurate.

During these traumatic years, Hu continued his journalism. In January 1932, he sent to a dozen friends in Beijing a proposal for a weekly review called *Independent Critic*. Ten people signed up; each made a financial contribution. "We are making *Independent Critic* because we all want to maintain always an independent spirit," Hu wrote. "No matter what party affiliation or religious belief, we will publish responsible comment to express the opinion of everyone. This is the spirit of independence."

Independent Critic's first issue came out in May 1932, with Hu as editor. It published articles on politics and current affairs, by its founding members and other contributors. Its circulation reached 13,000 at its peak. In the dangerous and highly charged atmosphere, it was hard to maintain independence. The country was facing a Japanese attack and was divided between the Nationalists and the Communists. All three forces wished to control the news and the 'narrative'.

Editing the magazine added to Hu's already heavy workload. The last day for contributions was Monday. He wrote in his diary of April 9, 1934 (a Monday): "In recent months, I was responsible for all the editing. I always worked until 03:00 to make the deadline. Dongxiu always scolds me and told me to stop the magazine. I say to her: "We have reached this moment. To sacrifice one day a week for the nation, what is that?'"

Over its five-year life, *Independent Critic* published 244 editions with more than 1,300 articles, of which more than half came from outside contributors. It was forced to close in July 1937, after the Japanese capture of Beijing.

"Dr Half-Finished"

Since Hu's arrival at BU in 1917, education in China had greatly expanded. A new school system was introduced in 1922; by 1931, there were more than 70 universities, up from 13 in 1921. But, despite the unification of the country in 1928, BU was in a precarious financial situation; professors received meagre salaries and books and equipment were in shortage.

In December 1930, Chiang Monlin, Hu's friend and fellow Columbia graduate, resigned as Minister of Education and took over as president of the university. He made wide-ranging reforms. He set up a university council as the main decision-making body of the institution, as well as administrative and provost councils.

In June 1932, he established three Institutes – of Liberal Arts, Science and Law. Hu became Dean of the Liberal Arts Institute. Each Institute had 14 departments. Hu's included Philosophy, History, Chinese Literature, Foreign Languages and Education. Hu's job involved teaching, as well as administrative and managerial duties. Between 1931 and 1937, Hu's focus was to work with Chiang to create the world-class university he had always dreamt of. By 1935, Beijing University offered 288 courses.

Such sweeping reforms were in large part made possible by an agreement in 1931 between BU and the China Foundation for the Promotion of Education and Culture (CF) to provide jointly two million silver yuan to the university over the 1931-35 period. Hu was a director of the CF. This money played a key role in promoting teaching and research.

In 1935, the university built a new library with space for 300,000 books and 500 readers. It hired leading academics and invited well-known foreign professors to lecture. By 1935, it had more than 40 research laboratories and 6,716 pieces of research equipment, making it one of the best equipped in the country.

Hu was a star lecturer, attracting 200-300 students to his classes. Even

In January 1932, during a visit to Shanghai, an American artist makes a sculpture of Hu Shih.

more attended the talks he gave elsewhere. Among young people, he was a hero, someone willing to challenge the government in public. People called this the "Hu Shih Cult". In January 1935, he went to Guangya Academy in Guangzhou to make a speech, but it was banned by the provincial governor. Instead, the 700-800 middle school students there surrounded and followed him off campus to hear the lecture.

In Beijing, he had a gruelling schedule. Up at 07:00, he left home for the university at 07:40. He returned for lunch at home, followed by a one-hour nap. At 13:40, he went to the office of the China Foundation (CF). He left there at 18:00, had dinner outside and met friends before returning home at 23:00: then three hours of work in his study until 02:00, when he went to sleep.

So, each night he had just five hours of sleep. There was almost no overlap with the schedule of his wife. In the morning, she attended to the house; at 14:00, she went to the house of friends for mahjong. She returned by car at 22:00.

Sunday was the exception. Hu held open house; anyone was welcome to come in and he received them. In the afternoon, he returned to his study and did not see visitors. In the evening, he also went out for dinner. The happiest time of the day was the five hours in the evenings, eating with and enjoying time with his friends. A very sociable person, he treasured these friendships. This was a comment his wife often made about the study where Hu spent much of his time: "In his room, there is very little space for the living but a lot of space for the dead. These books are all things left by dead people."

The house where the family lived, in 4 Xiamiliang ku, was among the most comfortable in Beijing. It had a spacious courtyard with 95 pine trees. It had three floors, including Hu's study and a large library that had once been a ballroom. The two boys had their own rooms; there were two bathrooms upstairs. The servants slept on the third floor – Madame Yang, manager of the house, a cook, two cleaners, a maid and a driver.

From 1930 to 1933, Hu used a noisy, second-hand car. In 1933, he paid US$1,090 for a Ford Deluxe Tudor Sedan; it arrived from the United

States on December 29 that year. Cars were an object of wonder in the city – there were only 2,000 on the road in 1937.

Hu could afford this lifestyle because of his high income. Each month the rent on the house was 80 silver yuan (US$18.1) and his BU salary 500 silver yuan. In addition, he received royalties of 15 per cent from sales of his many books, plus fees from articles he wrote in journals and magazine. This doubled his income to about 1,000 yuan a month.

A colleague of Hu's described the ambience in the courtyard home. Wen Yuan-ning obtained a law degree from Kings College, Cambridge University and then taught English language and literature at BU.

"Hu liked people and was good-hearted," Wen wrote. "On Sunday mornings, it was open house at Miliangku. No matter who came – student, Communist youth, businessman from Anhui or a beggar – everyone was welcome. He would certainly help the beggar. If someone was conceited, he would give them a lesson. If someone wanted a book, he would lend one. If it was a student, he would teach him something. Even if it was a windbag, he would give them a few proverbs. It was only in the evening, when things were quiet, that he could research a text or write his diary. It is because of this he became 'half-volume author'."

Wen said that Hu's friendly nature caused him to receive too many invitations and made him the best 'half-volume author'. Many friends made the same criticism, saying that he spread his time and interests too broadly, earning the nickname "Doctor Half-Finished". His eclectic interests included Chinese and Western literature, philosophy, politics, Buddhism and world affairs. He wrote the first volumes of the *An Outline of the History of Chinese Philosophy* and the *History of Vernacular Language* but never completed the second or third ones. "All these interests meant that he could not specialise," said Wen. "In addition, he was extremely cautious in what he wrote and would not publish anything lightly."

In 1931, a scholar named Feng You-lan published Volume One of the *History of Chinese Philosophy*, followed by a second volume in 1934. The two covered 1,041 pages, and 1,238 pages in English when translated by Derk Bodde. The first begin with the origin of philosophy in China and the second ends in the Republican period.

Four years younger than Hu, Feng followed Hu to the United States on a Boxer Indemnity Scholarship and earned his Ph. D at Columbia in 1923; like Hu, he studied under Professor John Dewey. Both men learnt the tools of western critical scholarship. After returning to China, Feng taught philosophy at major universities; he was chair of the Department of Philosophy at Tsinghua University from 1934 to 1938. His two-volume work became the standard work in the field of Chinese philosophy, eclipsing that of Hu's.

Hu continued his research in this area; based on it, he wrote articles, but not a second or third volume of his book. Feng was a scholar who dedicated himself to one field and became the national authority in it; Hu could have done the same, but chose not to.

Wen himself went on to a distinguished career as a writer, professor and diplomat. He served as ambassador of the Republic of China in Greece from 1946 to 1968 and then taught at National Taiwan University and other universities in Taiwan.

Student anger, Communist campaign

The Japanese takeover of Manchuria intensified student anger and protests across China. In December 1931, more than 200 BU students went to Nanjing and demanded that the government cut diplomatic relations with Japan.

On December 15, 600 students ransacked the Foreign Ministry building and attacked Cai Yuan-pei, the former president of BU, who had come out to talk with them; they beat him with sticks and left him bleeding.

While Hu shared the anger of his students against Japan, he disagreed with the way they expressed it. He opposed students staying away from classes. Better than they, he knew the severe imbalance between the military and industrial power of China and Japan. Attacking government institutions and beating up Chinese professors were not the way to deal with the threat.

Tensions between him and the students worsened after 1932; many were radicalised and called for war with Japan. In December 1935, the Communist Party in Beijing launched an anti-Japanese campaign among the students in the city.

One motive was to end the military campaign of the Nationalist army against the Communist base, numbering about 30,000 people, in the northwest province of Shaanxi. The army had surrounded the base and was closing in. If China fought a way with Japan, the army would have to end this siege. So, the party in Beijing mobilised students to boycott classes and demand action against Japan.

On December 9, the students marched. The Beijing police locked the city gates and turned water hoses on the students in the freezing weather. A week later, more than 30,000 marched in another demonstration in the city, as did thousands in Nanjing and other cities.

Some students denounced Hu for supporting the Tanggu Treaty. He received a threatening letter dated December 10 and signed by "A person who will kill you". In the toxic atmosphere of the time, this was a credible threat. Hu was angry that many students allowed themselves to be manipulated by rumours. He defended his position at public meetings; he also received letters of support from students.

Despite this fierce anti-Japanese atmosphere, Hu maintained friendships with Japanese scholars. He knew that the militarism of the Kwantung army represented only a part of Japanese society and that many opposed it. A photograph of June 1934 shows him meeting Daisetsu Suzuki. Suzuki was one of Japan's leading scholars of Buddhism, especially Zen. In 1933, he translated the *Lankāvatāra-sūtra* into English; this is an important sutra of Mahayana Buddhism, widely used in Chan Buddhism and Japanese Zen.

In 1963, Suzuki was nominated for the Nobel Prize for Literature. Hu and Suzuki would work together for over 30 years in discovering and editing original materials related to the early history of Zen Buddhism. Hu wrote extensively on this subject.

In January 1935, the University of Hong Kong invited Hu to receive an honorary doctorate. It was the first of 35 he would receive in his life, the most of any Chinese in history. In 1932, he had an operation on his appendix in Beijing.

Love Consummated

In 1933, Hu spent the months of July to September in Canada and the United States. It was his first visit to North America for six years. He had a crowded schedule – a five-day conference of the Institute of Pacific

Relations in Banff, Canada and six Haskell lectures in Chicago on "Cultural Trends in China" from July 12-24.

The World Fair was being held in the city at that time; a picture shows him being welcomed there at the Chinese pavilion. In late September, he gave a lecture at Vassar College in Poughkleepsie, New York. It was the alma mater of his friend Sophie Chen; she returned to the campus for his visit and accompanied him for the lecture. The large student hall was packed.

He was in demand wherever he went, including Honolulu, where he stayed for half a day on his way home. He gave a lecture on "A Philosophy of Life" at the University of Hawaii. Again, the hall was packed and many stood outside the windows.

He also went to New York, where he met friends, Chinese and foreign. They included his mentor Professor John Dewey; he found him "so superbly in good health and vigour. So humane and happy", according to a letter he wrote on September 15.

This gruelling itinerary was evidence of Hu's reputation in the United States and how it had endured despite a six-year absence. It also tells us that he had maintained contact with a wide network of people in North America despite his busy schedule in Beijing. He was the most famous Chinese in the United States.

Another milestone in his life occurred during this visit – the consummation of his love with Miss Clifford Williams, for the first time in 20 years of friendship. He was 42 and she 48. In April 1932, her mother had died, leaving her the large family home in Ithaca, New York. To keep the house running, she accepted boarders. With the passing of her mother, she became the mistress of the house.

In 1935, Hu Shih receives an honorary doctorate at The University of Hong Kong.

During the visit to North America, Hu went twice to Ithaca, once for two weeks in August and the second time for 24 hours. We owe knowledge of this episode to Professor Chou Chih-p'ing, Professor of East Asian Studies at Princeton University.

In 1997, Chou was doing research in the library of Beijing University and inquired if they had the letters which Miss Williams had written to Hu Shih during their friendship of 40 years. The staff handed him the letters, including those for 1933. They had survived the anti-Hu campaigns of the 1950s only because they were written in English; few at that time knew of their significance.

His letters to her during this period, on the other hand, were retyped by her with omissions and her own summaries; a shy and self-effacing person, she did not want the outside world to know of her affair with a married man nor to cause pain to his wife. So it is from her letters that we feel the passion and intensity of those two visits in 1933. With Susan Chan Egan, Professor Chou published in 2019 *A Pragmatist and a Free Spirit, the Half-Century Romance of Hu Shi and Edith Clifford Williams.* Their excellent book has been one of the main sources for this biography; we thank them very much.

Since Hu had returned to China in 1917, Miss Williams had lived in the family home in Ithaca, working at the library in Cornell University and caring for her parents and other family members. She did not marry. During those 16 years, she and Hu continued their correspondence despite his marriage and busy schedule. With the passing of her mother, she could behave as she wished in her own home; also, perhaps, both felt the passage of time – with a war likely in China, when would the two be able to see each other again?

This is what she wrote in a letter to him on September 13: "Motives and reasons dwindle to nothingness beside this great thing between us – seeming so much greater than either of us – intangible as the sun (whose heat is burning). I miss your body but I miss more terribly your total presence, the bit of you wherein I live. What is you in me is always craving the me in you."

On September 22: "Were it not better never to meet than to know parting? All my life I have ascetically inclined to renounce entirely rather than take a little of something. Why can we not take this possible day, simply, as a gift and enjoy companionship during the few hours it holds, adding them in memory to the few we have?"

On September 26: "Hu Shih, I love you! ... I am very humble that you should love me – but, at times, your love surrounded my thoughts like sunlit air ... If there is chance for us to be together living fully, can you imagine our not merging like two streams long seeking the same valley?"

On September 27: "The jacket of formality has slipped to the floor – you know me utterly now, Hu Shih – would you prefer the lady of illusion? She may have been a fine thing but it is I, of the small breasts and inadequate housekeeping, of the fuzzy, inadequate brain, who have touched your body and eyes. I cannot believe you can love so poor a thing, yet your love wraps me round."

As Hu crossed the country West for his boat home, the house at Ithaca became very empty. "Loneliness stalks and shakes us terribly by the shoulders," she wrote on October 1. "After you go back, do not try to write ... I know the flood of work you will be absorbed in. Letters are difficult. You have given so much! Nothing should be demanded."

Her intellectual self told her that, because Hu would not leave his wife and career in China, they could never marry. But their two weeks together gave her a glimpse of what that marriage might have been – deep passion and deep pain. Since she edited his letters and removed certain passages, we do not know if Hu's letters to her expressed sentiments of similar intensity. After his death in 1962, she donated the letters she received from him to the Hu Shih Memorial Hall in Nangang, Taipei.

Nightmare begins

At dawn on December 12, 1936 troops under Zhang Xue-liang, the former ruler of Manchuria, stormed the headquarters outside Xian where President Chiang Kai-shek was staying. They killed most of his bodyguards. In his night clothes, Chiang escaped from the house, climbed over a wall at the back of his compound and hid in a cave on the side of a mountain. Zhang's soldiers captured him there. Zhang gave eight demands to his hostage; the most important was to end the campaign against the Communists and present a united front to Japan.

After two weeks of complex negotiations, Chiang flew back on Christmas Day to Nanjing, where he was given a rapturous reception by a crowd of 400,000 people. The Communists had succeeded in their objective of lifting the siege against them. Unwisely, Zhang chose to fly on the same plane as Chiang. On arrival in Nanjing, he was detained and kept under house arrest for the next 53 years, in the mainland and Taiwan – making him the world's longest serving political prisoner. This kidnapping came to be known as the "Xian incident".

The outcome was bad news for Japan, which wanted a divided China and a government that would collaborate with it. So, its generals accelerated their plans for war. One reason to proceed quickly was the absence of any move by the United States, Britain or other Western powers to come to China's aid and attack; Japan enjoyed military superiority on land, sea and air. The European countries were absorbed in facing the threat of Nazi Germany.

In a letter on September 13, 1936 to Miss Williams, Hu wrote: "I have been doing all I can to avoid a war. But I have now very grave doubts as to the slightest possibility of peace if your country does not take an active part in the international affairs of the Pacific." The phrase he used was

"Peace is more difficult than war."

On July 7, 1937, the nightmare everyone had feared came to pass. The truce between Chinese and Japanese soldiers in the Tianjin-Beijing theatre that had lasted since May 1933 finally ended; fierce fighting erupted between the two sides in Wanping, a walled city 16 kilometres southwest of Beijing.

It was the beginning of Japan's all-out attack on China and the start of World War Two. That day Hu was attending an education conference in Nanjing. He was devastated; it meant that all hopes of a negotiated agreement with Japan had failed. He accepted the idea that China had no choice but to make total war and that he must play a role in the war effort.

A week after the war broke out, President Chiang convened a conference in Lushan, Jiangxi province ; he invited leaders from all sectors of society. Hu was present, together with other university chiefs. Chiang wanted to lay out the national strategy for a war that he knew would last a long time and devastate the country, and to rally everyone behind this strategy.

It was indeed a historic moment. In a history of thousands of years, China had never faced an enemy with the destructive power of the Imperial Japanese Army. Those listening in the large conference hall at Lushan knew that many of those present would not survive the war. On July 17, Chiang made the keynote speech, entitled "The Limit of China's Endurance". Its apocalyptic nature matched the gravity of the situation:

"The consequences of this incident (on July 7) threaten not only the very existence of China, but the peace and prosperity of mankind ... In its policy toward Japan, the National Government has constantly followed these principles (of peace), in the hope that the confusion caused by

Japan's arbitrary actions might be overcome, and all problems might by dealt with through recognised diplomatic channels.

"Since we are a weak country, there is only one thing to do when we reach the limit of endurance: we must throw every ounce of energy into the struggle for our national existence and independence. When that is done, neither time nor circumstances will permit our stopping midway to seek peace ... Only a determination to sacrifice ourselves to the uttermost can bring up ultimate victory ... We cannot do otherwise than resist and be prepared for the supreme sacrifice. We have not sought war. It will have been forced upon us ... Once the battle is joined, there can be no distinction between North and South, nor between young and old. Everyone will have to give everything that he has ... When you return to your home districts, I trust that you will pass this message to all the people, so that they may understand clearly the present situation and be absolutely loyal to the state." Chiang's words accurately laid out what would happen for the next eight years.

Sitting in the front of the hall was former Prime Minister Wang Jing-wei. Like Chiang, he knew well that China was militarily weak; but he drew a different conclusion. In December 1938, he left China for Hanoi and began negotiations with Japan. In March 1940, he established a pro-Japanese government in Nanjing.

By July 20, the Japanese army had more than 180,000 soldiers in the Tianjin-Beijing area. The Chinese army defended valiantly against the superior firepower and fighter planes of their opponents, but in vain. On July 29, the Japanese army captured Beijing.

Before it arrived, Hu's wife left the city with youngest son Si-du for the safety of the French concession in Shanghai; they took with them 70 crates of Hu's books, manuscripts and notebooks. The elder son Zu-wang was studying in the interior, and Hu himself was in Nanjing.

In September, on the orders of the Ministry of Education, Hu's colleagues and students of Beijing University left the city and moved to Changsha, capital of Hunan province, where they formed a new university with two other institutions. In December 1937, the Japanese intensified their bombing of Changsha, so the university moved again. In February 1938, the faculty and staff started a 68-day journey across Hunan, Guizhou and Yunnan provinces to Kunming; they travelled 1,600 kilometres, of which 1,300 was by foot. From April 1938, BU became part of the National Southwestern Associated University in Kunming, where it remained for the rest of the war.

Chiang asked Hu to go abroad to mobilise support for China's war effort. Initially, Hu was reluctant. He preferred to remain in Nanjing and share the dangers of the war with his friends and colleagues. What would they think of him enjoying the comfort and safety of North America as they ran to underground bunkers to escape Japanese bombing? Then President Chiang asked him: "Should we abandon diplomacy? Should we not ask the democratic countries to support us in the war?"

This was undoubtedly the best use of his talents. He could do much more for his country abroad than at home. His eloquence, global reputation and network of high-level contacts would enable him to reach a wider audience than any other Chinese. And, like Chiang, he knew that China could not win the war on its own. So, he accepted the offer. His condition was that he went in an unofficial capacity and would speak on his own behalf. He did not want to be restricted by the orders and regulations of the government.

Sources for Chapter Six

A Pragmatist and his Free Spirit, the Half-Century Romance of Hu Shih and Edith Clifford Williams, by Susan Chan Egan and Chou Chih-p'ing, Chinese University Press, Hong Kong, 2009.

Beijing University website

Hu Shi and His Friends, 1904-1948, Commercial Press Hong Kong, 1999.

Article in Peking University Press, "Why are so many of Hu Shih's Works Half-Complete?", published on 20/1/2019.

If Not Me, Then Who? Hu Shih Volume Three, Chiang Yung-chen, Linking Publishing Company, Taiwan, first edition January 2011.

Retracing the Journey of Hu Shih, A Re-examination of his Life and Thinking, by Yu Ying-shi, Linking Books, Taipei, 2014.

Diary of Hu Shih

The Search for Modern China, by Jonathan Spence, W.W. Norton & Company, 1990.

Taiwan Review, "The War of Resistance 1937-1945", published on 1/7/1987.

Selected Works of Hu Shih, Wang Ching-tian, National Library Publishing Company, New Taipei City, 2020.

The Collected Wartime Messages of Generalissimo Chiang Kai-shek 1937-1945, Volume One, Internet Archives.

CHAPTER SEVEN

SAVING CHINA

On the morning of September 26 1937, Hu arrived by plane in San Francisco, having left Nanjing on September 8. It was supposed to be a journey of about two months in North America. He did not imagine that, a year later, he would become China's ambassador to the United States, and his absence from home would last eight-and-a-half years.

Initially, he was reluctant to go because he felt uncomfortable in an official role. Throughout his life, he had been a free and independent spirit, able to say and write what he wished. It was this independence that gave his words power; everyone knew he was not speaking on behalf of a political party or a corrupt warlord. In addition, he felt guilty to leave his family, friends and colleagues in such danger behind. The Japanese military controlled the air and was bombing Nanjing, Changsha, Wuhan and other cities.

In a letter to Miss Williams on 25/9/1937, Hu wrote: "I must say that it was much against my own wishes that I should leave China in the midst of daily and hourly perils and live in comparative comfort and complete safety in foreign lands ... I decided to come on the condition that I do not carry any diplomatic mission, not be required to do 'propaganda' work. I am here merely to answer questions, to clear up misunderstandings, and to present my own viewpoints. My itinerary is still uncertain."

Hu's mission was to turn the sympathy of Americans for China into concrete aid and assistance, especially military and financial. He was pessimistic – his knowledge and experience of the U.S. told him that, while sympathetic to China and the terrible sufferings of its people, the American government and people did not want to enter a war. It was far from their shores and did not directly affect their interests – other than a limited number of American companies and several thousand American residents in China.

So, he believed that, despite his charm, eloquence and excellent contacts among the American elite, it would be mission impossible. In 1935, 1936 and 1937, the U.S. Congress passed Neutrality Acts aimed at ensuring that the country did not become involved in foreign wars. This reflected widespread disillusion with the failure of the major powers after World War One to create a stable world order.

The U.S. Senate vetoed participation in the League of Nations, so the country never joined. Nor did it ratify the Peace Treaty of Versailles. Two Acts, in 1921 and 1924, limited overall immigration and set country-specific quotas, which privileged migrants from northern and western Europe. The number of arrivals fell to less than 20 per cent of the pre-World War One totals. Most Americans were descendants of those who had left Europe to escape its national and religious wars, poverty and religious and ethnic persecution. Why should they help those who still engaged in such conflicts? Many believed the United States was "God's country" precisely because He had enabled them to escape from Europe's poisonous history.

The Neutrality Acts of 1935 and 1937 prohibited exporting arms and ammunition to any foreign nation at war. The 1937 Act banned Americans from travelling on ships owned by any belligerent nation and declared that American-owned ships could not carry any arms intended for war zones. A poll in 1938 found that 70 per cent of Americans believed their country's participation in World War One had been a mistake. In 1938, Maxwell Hamilton, Chief of the Far Eastern Bureau of the State Department, told Hu: "This is a matter of survival for China. It must fight for itself. No-one else can help her."

In the face of such hostility, how could Hu Shih persuade the U.S. Congress and public to come to the aid of a country with which they had no historic nor racial connection?

On the afternoon of September 26, the day of his arrival, he gave his first talk in San Francisco. It was at the Chinese Theatre; the title was "Make the Best Calculation, Work to the Uppermost". Three days later, he addressed an audience in the same city at the Commonwealth Club on "Can China Win the War?". He said justice was on the Chinese side and victory was only possible if everyone stood together against the Japanese threat. On September 30, he gave a lunch speech at the University of California. Then he gave a series of nationwide radio talks on the Columbia Broadcasting System.

He had an exhausting schedule. In the 51-day period between January 24 and March 16 1938, he gave 56 speeches, 38 in the U.S. and 18 in Canada. He spoke at universities, public institutions and civic associations. His audiences included members of Congress, business leaders, religious communities, the media, lawyers and academics as well as the general public. He crossed the American continent from west to east. His reputation and eloquence and widespread public interest in the war ensured large audiences for his talks.

The mainstream U.S. media gave broad coverage to the conflict; it favoured the Chinese side. It was clear who had started the war and who had tried to prevent it. The bombing of civilian targets and atrocities committed by Japanese soldiers in Nanjing and other places enraged the American public. All this led to a favourable hearing. Hu knew well that, in the U.S. more than in most countries in the world, public opinion played a major role in shaping government policy.

But the economic interests of the U.S. favoured Japan over China. Japan ranked second, after Britain, as the biggest market for American exports;

China was only in seventh place. U.S. trade with Japan was more than twice that with China and its capital investment there was limited. Since Japan had almost no natural resources, its arms industry depended on foreign countries. In 1937, it imported 2.4 million tonnes of scrap iron, of which 1.78 million, or 73 per cent, came from the United States. The U.S. was also its largest supplier of petroleum; in 1939, it provided two thirds of its petroleum imports. These were strong economic interests to stay out of the war and allow this lucrative trade to continue.

The U.S. was a large and complex country, with many disparate and competing interest groups. How could one person persuade the government to intervene in a war 10,000 kilometres away in which the country had no vital interests?

Since Hu had left his motherland, the military situation had greatly deteriorated and hundreds of thousands of Chinese had died. The Battle of Shanghai lasted from mid-August until the end of November. The Chinese toll was 187,000 soldiers killed and 83,500 wounded in weeks of savage fighting, while the Japanese side lost 59,000. It led to the capture of Shanghai, outside the foreign concessions; often called "Stalingrad on the Yangtze", this was one of the biggest battles of World War Two.

After this victory, the Japanese military marched west and captured Nanjing, the national capital, on December 13. There its troops committed terrible massacres and rapes; the Nanjing War Crimes Tribunal in 1947 put the number of dead at more than 300,000.

Chiang moved his capital to Chongqing in the remote southwest and took many of his best troops; he wanted to save them for future battles. As Hu journeyed across the United States, he read of these tragedies with horror and sadness. They put even more pressure upon him to persuade the United States to help his devastated country.

In late August 1938, he learnt of the death of one of his closest friends, Hsu Sing-loh; on August 24, the Japanese shot down the civilian plane on which he was flying from Hong Kong to Chongqing. He was just 48. After studying in Britain and France, Hsu returned to China in 1914 and held senior economic posts in the government and private banking. He also taught in the economics faculty of Beijing University. If Hu had stayed in China, he could have been on that plane. In a letter to Miss Williams on August 25, 1938, he said that Hsu "was my best friend. The most learned banker in China and the finest man that can be found anywhere ... I am really in great sorrow."

Hu's best hope was Franklin Roosevelt, President since 1933. He opposed the Neutrality Act, which limited his freedom to support friendly nations; and he backed China in the war. In a speech in Chicago on October 5, 1937, Roosevelt proposed the use of economic pressure against "lawless" nations.

In Washington, Hu met Roosevelt, in the company of China's ambassador Wang Cheng-ting. The president asked him if the Chinese army could survive the winter. Hu replied: "China needs the support of the United States. We believe His Excellency the President will very quickly make the right judgement based on clear foresight!"

But Roosevelt was bound by the neutrality policy. When the two men left, he shook Hu warmly by the hand, to show his sincerity. Their close relationship would prove critical over the next four years. From the second half of 1937, Japanese warplanes started to bomb civilian areas of Chinese cities, including those where Americans lived. These raids caused outrage in the U.S. On July 1, 1938 the Department of State told aircraft manufacturers and exporters that the U.S. government was strongly opposed to the sale of aeroplanes and aeronautical equipment to Japan.

On December 13, the American Foreign Policy Association hosted a forum at the Waldorf Astoria, one of the most famous hotels in New York. In attendance were members of the city's elite and journalists, including those from China and Japan. It was the same day that the Japanese army entered Nanjing.

Hu made an emotional appeal to the audience about how China was fighting alone in an existential struggle for its survival. He described the atrocities committed by the Japanese military against civilians. This produced an angry response from two senior Japanese editors in the room, from the *Osaka Mainichi* and *Tokyo Nichinichi*. The audience was on the edge of their seats; they were witnessing the war being played out in front of them.

On December 20, Hu wrote a letter to Miss Williams wondering how useful all his talks had been. "Really there is very little I can do. The incidents I had long predicted at last came. They are not yet settled. This is the beginning of a new phase in the war – an international phase which is fast overshadowing what is happening inside China … I long for the time when I can get back to work and for which I have prepared myself all these 20 years. Now that grey hairs are on me, I find I must not waste myself any longer. What a waste for me to talk war and international politics!"

In the spring of 1938, Harvard and the University of California invited him to a teaching post in the next academic year. Friends advised him to accept one of the offers. But he declined in light of the hardships being endured by his colleagues and students at home. During his time in the U.S., he renewed his friendship with Miss Williams. Together they took a voyage of 700 miles in her car, probably to Canada. Their feelings for each other were as intense as before.

At 52, she had received an offer of marriage from a man who was probably a professor at Cornell University. She described him as patience, courteous and gentle but unmusical, poorly read and lacking in imagination and slow in vision. She declined. "I cannot marry all those who want to marry me!" she wrote to Hu on November 11. "And ironically neither can I marry the only man I ever wanted." In the end, she never married anyone.

Becoming Ambassador

In July 1938, Hu went to the United Kingdom and on to Paris. There he received a cable from President Chiang inviting him to become China's ambassador in Washington. Two of his friends, Wellington Koo, ambassador in Paris, and Kuo Tai-chi, ambassador in London, as well as Prime Minister H.H. Kung, sent cables urging him to accept. His initial response was to refuse; he had been a free spirit for more than 20 years and did not want to accept the restraints of working in a government which he would have to obey. But his friends disagreed, saying that, in this critical moment in his country's history, he could not shirk his responsibility.

The U.S. was the only country that could save China; it alone had the military power and naval strength to defeat Japan. On its own, China could not win the war; it could only prolong it, at terrible cost to its people, its cities, towns and villages. There was no other Chinese with Hu's knowledge, eloquence in English and connections among the U.S elite. This was the moment that history had prepared for him. On July 26, he finally decided to accept. This is what he wrote in his diary: "The nation is in such peril. Facing such an order, how can I refuse?.

On September 17, the government announced that Wang Cheng-ting had resigned as ambassador and that Hu Shih had replaced him. This is how *The New York Times* reported the news: "Few Chinese are so thoroughly representative of the best of the new and old China ... few are so well qualified to explain China to the United States and the United States to China."

Hu presented his credentials to President Roosevelt on October 28. After the presentation, he gave a news conference at the embassy. In a letter to Miss Williams after his appointment, Hu wrote: "I am seriously undertaking my present task in the belief that I may grow more fit for it. At least I do not like it now. It may be that the fight necessary in the work may make me like it more and more. I hope so. I have 'degenerated into an ambassador.'"

He went on to develop strong personal relations with Roosevelt and his wife Eleanor, Secretary of State Cordell Hull and Secretary of Treasury Henry Morgenthau. Hull described him as "One of the ablest and most effective public servants this government has had in the foreign diplomatic corps in Washington."

He also enjoyed personal connections with prominent journalists and people in the legal profession. As a result, he and China's cause received excellent coverage in the media.

Hu had to decide whether to invite his wife to join him. The spouse of an ambassador plays an important role as hostess at social events and building her own personal network. Dongxiu had not seen Hu since he had left China in September 1937; initially, his absence was only for several months. Hu decided that his wife was not suited to be the lady ambassador. She did not speak English and had no interest in the matters that were discussed at embassy functions. Her main activity was mahjong; many Americans would consider it a form of gambling.

Instead, Hu used the wife of the Embassy Secretary as hostess for his receptions; highly educated, she was well suited to this role. For staff, he hired a Belgian as a butler and five European refugees as servants. This was most unusual. For security reasons, embassies usually hire staff from their own countries. He wrote regular letters to his wife and sent her money and gifts.

The new position brought more prominence and media coverage. On December 4, 1938, he gave a speech at the Harmonie Club in New York: "I would not hesitate to say that China is literally bleeding to death ... We have suffered one million casualties ... 60 million civilian sufferers ... (are) fleeing the invader and roving the country without shelter, without medical aid and in most cases without the barest means of subsistence ... China is now entirely cut off from all access to the sea."

He compared China's struggle to the American army at Valley Forge in the Revolutionary War in the winter of 1777-78. On February 6, 1778, France and the United States signed a treaty to create a military alliance. A major reason why the Americans won their independence, Hu said, was this support from the French. The Chinese were like the Americans at Valley Forge.

On the night after the speech, he suffered a serious heart attack. He was treated at the Presbyterian Hospital in New York. In a letter to Miss Williams on February 1, he wrote: "I had a heart attack, which I thought was a mere acute indigestion ... It is a timely warning to me just as I

At a reception at the Chinese embassy in Washington, Hu Shih talks to Eleanor Roosevelt, wife of U.S. President Franklin Roosevelt.

begin to enter my 'middle age'. I shall have to stop smoking entirely and to reorganize my life radically."

It was so serious that he was confined to the hospital for 77 days before the doctors allowed him to return to Washington. During that time, he did not write a single diary entry. He had to behave carefully; he could not climb many stairs. It was a reminder of the stress of a government post and why he had been so reluctant to accept it. During his four years as ambassador, he gave 246 talks, an average of more than one a week.

Disappointment

Back at his desk in Washington in February 1939, Hu resumed his busy diplomatic and social schedule. The next 19 months were full of disappointments. He lobbied legislators and journalists to revise the Neutrality Acts. But, for all the meetings with sympathetic officials, well-attended public speeches and positive media coverage, he failed to change U.S. policy.

At home, the situation was deteriorating. The Japanese had failed in their objective to achieve a quick victory and force President Chiang to sign a peace treaty. Instead, the war became bitter and long drawn-out, with rising military and civilian casualties on both sides. With their command of the air, the Japanese continued bombing raids on major cities, including Chongqing, the wartime capital. They aimed to kill President Chiang, his family and cabinet; but they lived in villas hidden in the mountains around the city and survived. Instead, the raids killed thousands of civilians. The residents built 1,700 tunnels to take shelter during the raids.

The Japanese pioneered attacks on civilian areas, a tactic used later in the war by the German, British and American air forces. During 1939, the Chinese army won four battles but failed in its first large-scale counter-offensive in early 1940. The Japanese controlled northeast China and major cities in the east and south. But it had not forced China to surrender – the war objective of the Imperial General Headquarters.

The puppet governments the Japanese installed in occupied areas were unpopular and ineffective. Guerilla resistance continued in these areas. In March and June 1940, Chinese and Japanese officials held secret talks, in Hong Kong and Macao, over a possible peace agreement. They came to nothing because the Chinese side refused official recognition of Manchukuo and the stationing of Japanese troops in North China.

The talks were a sign of Chiang's precarious position. In July 1940, under Japanese pressure, Britain closed the Burma Road – the land link between India and China. In September 1940, the Japanese occupied French Indochina, cutting the railway between Haiphong (in Vietnam) and Kunming. China had lost all land routes to the east and to the south.

But these tumultuous events did not touch the vital interests of the U.S.; moral outrage was not sufficient reason to stop trading with Japan, even less to intervene in the war. They had even less influence on the United Kingdom; it was locked in a life-and-death battle with the Germany of Adolf Hitler.

Hu was becoming dispirited. In a meeting in mid-1940 with Stanley Hornbeck, senior East Asian affairs advisor at the State Department, Hu said that the situation in China was deteriorating. "China needs from the U.S. more positive action than mere words of encouragement, something more positive than mere promises of financial assistance."

Then, on September 27 1940. Japan signed the Tripartite Pact with Germany and Italy. Article 3 of the Pact said: "Japan, Germany, and Italy agree to undertake to assist one another with all political, economic and military means if one of the Contracting Powers is attacked by a Power at present not involved in the European War or in the Japanese-Chinese conflict." This "Power" was clearly the United States.

For Hu, this was a game-changer. The pact convinced American officials that the conflicts in Europe and Asia were part of the same war. Clark Eichelberger, National Director of the Committee to Defend America by

Aiding the Allies, wrote in late 1940: "China in the Pacific, with Britain in the Atlantic, now constitute our first line of defense."

On November 5, Franklin Roosevelt was re-elected President of the United States; he carried 38 states, against 10 for his Republican rival Wendell Wilkie. To win, he had to promise no involvement in foreign wars. For Hu, it was a victory. He knew the President well and that his sympathies lay strongly with China and Britain.

On November 30, the U.S. approved a US$100-million credit for China. In early 1941, the head of the China branch of the Office of Strategic Services, predecessor of the CIA, took up his post in Chongqing. On March 11, 1941, Congress passed the Lend Lease Act, which gave Roosevelt the power to sell war materials to any government whose defence he considered vital to U.S. security. Initially, it was aimed at Britain; on May 6, it was extended to China.

But the U.S. continued to trade with Japan, including exports of the oil that enabled its tanks and aircraft to operate in China. On June 22, three million German troops invaded the Soviet Union. This was good news for China – Japan's ally had taken on a giant and powerful enemy. In the summer of 1941, U.S. volunteer airmen arrived in Chongqing, the start of the "Flying Tigers" operation which would play an important part in the war.

Unwelcome Visitor

But this progress was not enough for President Chiang in Chongqing. He believed Hu had not been effective enough in persuading the U.S. government; he wanted a more powerful lobbyist there. He chose his brother-in-law, T.V. Song. Like Hu, he was a graduate of Columbia

As ambassador in Washington, Hu Shih explains to U.S. President Franklin Roosevelt the contents of a declaration signed by thousands of Chinese.

University; he had served as Minister of Finance from 1928-1933. In personality, he was the opposite of Hu. One was soft-spoken and polite, the other loud and arrogant. Song arrived in Washington in June 1940. According to an entry in Hu's diary, the newcomer told him: "Many people in China say that you make too many speeches and do not attend to business. You had better pay more attention to your job."

The two men had a difficult relationship. Hu saw Song blundering through relationships he had spent years cultivating. Song's presence in Washington made Hu dislike the job even more. American officials asked who represented the Chinese government – its ambassador or the brother-in-law of the President?

"Hu loses temper with President Roosevelt"

During the first half of 1941, the United States repeatedly tried to stop Japan from moving into Southeast Asia. But, in July, 140,000 Japanese troops invaded southern Indochina. In retaliation, the United States imposed an embargo on oil sales to Japan – given its dependence on American oil, this was an economic declaration of war.

In the autumn, the two sides held negotiations; they became increasingly confrontational. Washington demanded that Japan withdraw its forces from China; Tokyo adamantly refused. An Imperial Conference in Tokyo on November 5 agreed that, if they could reach no satisfactory diplomatic solution by December 1, they would launch a war against the U.S. For the rest of November, diplomats of the two countries engaged in frantic negotiations to avert a war.

On November 24, the U.S. government proposed a three-month "modus vivendi". Clause Three said that Japan would withdraw forces in southern French Indochina and reduce the number of troops in (all) French Indochina to no more 25,000. Clause Four allowed the resumption of U.S. exports to Japan, subject to conditions, with petroleum only for civilian needs and on a monthly basis. Clause Seven dealt with China – but made no demand for Japan to withdraw its troops.

Shown the draft, the British, Dutch and Australian representatives in Washington gave their approval. It would have stopped the attack on Pearl Harbour – the Japanese battle fleet set out for Hawaii on November 25 – and averted a war between the two strongest military powers in the Pacific, at least for a few months. But when he was shown the draft, Hu was enraged.

On November 26, Hu met President Roosevelt and expressed his intense opposition. It was the most important meeting of his diplomatic career and, probably, of his entire life. The U.S. was the only power in the world capable of defeating Japan; China could not. If Washington made peace with Tokyo, Japanese troops would be stationed in China for the long term and Manchuria would probably be lost forever.

According to a report in *LIFE* magazine on December 15 that year, "For the first time in his life, the soft-spoken scholar (Hu) is reported to have lost his temper ... He reminded the President of his many, freely given pledges to China." He convinced Roosevelt to veto the modus vivendi. The President summoned the Japanese representatives to the White House to tell them. "Hu played a crucial role in preventing an agreement between the U.S. and Japan," *LIFE* reported.

With no agreement by December 1, as the Imperial Conference in Tokyo had ruled, the Japanese battle fleet continued its journey toward Hawaii. Just before 08:00 on December 7,350 aircraft bombed the U.S. fleet in Pearl Harbour. They sunk four battleships and damaged the other four and killed 2,400 Americans.

Hu was the last diplomat with whom Roosevelt conferred on the morning of December 7, 1941. He went back to the embassy and was immediately called to the phone. "Hu Shih, the Japs have bombed Pearl Harbour. I want you to be among the first to know." These meetings show the strong personal bond between the two men which Hu had created over many months. Would Roosevelt have given an audience to another Chinese ambassador that November 26? If he had, would he have been persuaded by him?

The attack decided the outcome of the war in Asia. While it was a military success for Japan, it was a political, moral and strategic disaster. The fact of being a "sneak attack" and the number of American dead enraged the American people and gave Roosevelt the wide popular support he needed to declare war.

The Japanese admiral who led the attack, Isoroku Yamamoto, knew that Japan would lose a war against the United States. He had opposed the occupation of Manchuria, the war in China and the Tripartite Pact. He had studied at Harvard University and served two postings as naval attaché at the embassy in Washington; he could speak fluent English. Yamamoto had travelled extensively in the U.S. and knew that its production capacity of iron and steel, aircraft, armaments, petroleum and everything else needed for war far surpassed that of Japan. Asked by his Prime Minister in mid-1941 about a possible war with the U.S., he replied: "I shall run wild considerably for the first six months or a year, but I have utterly no confidence for the second and third years."

This is exactly how it turned out. After its defeat at the Battle of Midway in June 1942, Japan began to lose the war. If it had chosen to attack only British and Dutch colonies in East and Southeast Asia, the U.S. might not have entered the Pacific War. So Hu's friendship with Roosevelt and his intervention on November 26 played a critical role.

Bitter Taste

Immediately after Pearl Harbour, President Chiang appointed T.V. Song Foreign Minister and ordered him to remain in Washington, to manage the country's relations with the U.S. and Britain. This was a highly unusual arrangement – and very damaging to Hu. From then on, Chiang could deal with the U.S. government directly through his brother-in-law,

After the Japanese attack on Pearl Harbour, President Franklin Roosevelt on December 22, 1941 signs a proclamation of 26 Allied nations at the White House. China is one of the four major nations. Hu Shih is holding a copy of the proclamation outside the White House.

sidelining the embassy and the ambassador.

As Foreign Minister, Song was also Hu's direct superior. He ordered the embassy staff to show him all communications from Chongqing – but did not show Hu what he received himself. Song's main work was to negotiate loans for China, something in which Hu was not involved.

Relations between the two men deteriorated. They had opposing styles of diplomacy. Song was aggressive and outspoken. The U.S. State Department preferred Hu; it described Song's behaviour as "arrogant, improper for foreign relations and stepping out of line."

Chongqing sent Hu US$60,000 to do propaganda. He sent it back, saying: "My speeches are sufficient propaganda and they do not cost anything. Propaganda is unnecessary for a diplomat accredited to a friendly government and people. His duty is to understand and appreciate the county to which he is accredited ... the rest is easy." In the Chinese pecking order, it was clear who was more senior.

Song often asked Chiang to replace Hu. He did so finally in September 1942 and appointed Dr Wei Tao-ming.His curriculum vitae did not suggest a candidate suited for such an important job – perhaps the very reason why Song wanted him.

Wei had spent six years obtaining a Ph.D. in law at the University of Paris; in 1925, he returned home and served in senior positions. In 1941, he was appointed ambassador to France but was unable to take up the post. So, he did not have Hu's detailed knowledge of the U.S. nor his fluency in English or personal network.

Cordell Hull praised Hu for his "outstanding contribution to Chinese-American friendship and the extraordinarily capable manner in which he had discharged his heavy responsibilities during his tenure." *The New York Times* called his dismissal a "mistake, unless some higher post is reserved from him at home".

Other newspaper editorials and personal letters expressed regret at his leaving his post. In a letter to Miss Clifford on September 24, he wrote: "The change came as a really great relief to me. I am ready for a few weeks of complete rest before making any definite plans for the future ... It is wonderful to be a free man and have leisure to sleep. Medical advice is against my flying back over very high altitude. Lack of books in China today is also a factor against my returning now."

Return or Stay

Hu had to decide whether to return home or stay in the United States. The Chinese government offered a new post of Higher Adviser to the Cabinet; major American universities proposed teaching positions. His wife was in China, their two sons in the U.S.. Zu-wang had just graduated from Cornell and Si-du was about to go to college there; this would require substantial expense.

In the end, Hu decided to stay in the United States; he would remain there for the next four years, returning to China only in July 1946. He declined the teaching positions but accepted a post as researcher and consultant on culture to the American Council of Learned Societies.

Based in New York, this council was created in 1919 with the mission of "advancement of humanistic studies in all fields of the humanities and social sciences and the maintenance and strengthening of national societies dedicated to those studies."

Two years later, he did limited teaching – at Harvard from October 1944 to June 1945 and Columbia University in the autumn of 1945. He also gave six lectures at Cornell in February 1946.

After four years in the maelstrom of Washington diplomacy and politics, he wanted peace and quiet and the freedom he enjoyed during the first 20 years of his professional life. He wanted to do his own research more than teaching; during the previous four years, he had given hundreds of public speeches and attended innumerable social events. He exhausted himself in the process.

There were other reasons for his decision. One was to spend more time with his sons; while they were growing up, he had seen little of them. Another was the need to earn the money to pay the college fees of Si-du; he would be unable to do that on a salary in China, which was suffering from rampant inflation. A third was that he was safe and comfortable in the U.S., with a wide network of friends and contacts. His health was fragile; less than four years before, he had had a life-threatening heart attack.

The fourth reason was to avoid the nest of vipers that Chongqing had become, with fierce conflicts between the government and its American advisers and bitter personal rivalries within the government. As a former ambassador to Washington, he would be unable to stay out of these conflicts. He had achieved the mission which President Chiang had entrusted him – to bring the U.S. into the war and ensure the defeat of Japan. What better service could he render his country?

Finally, there was nowhere in China with the conditions to enable him to do the kind of research he wished. With two other colleges, Beijing University had set up a new campus in Kunming, capital of Yunnan province, in the far southwest.

This is how the university website describes this period: "It was extremely difficult to run a school in wartime. The library bookshelves were made of stacks of wooden crates and only a few dozen thousand books were kept. The seats in the library could only seat a few, so some students had to frequent nearby teahouses – not just for reading and discussions, but also a glass of water. The 'NSAU (National Southwestern Association University) Students Teahouse' became a unique sight in wartime Kunming. The teachers would have to make do with makeshift, handmade laboratory devices. The ever-increasing prices made the lives of teachers and students increasingly difficult. From the beginning of the War of Resistance against Japanese Aggression to 1943, prices in Kunming rose by 300 times, but the salaries of NSAU staff rose only fivefold. To make ends meet, the teachers had to do part-time teaching at other schools or even sell their books and clothing at a low price. The students also had to take up part-time jobs between classes."

Kunming's infrastructure and military bases were under regular attack by Japanese bombers. Under such conditions, how could Hu do textual study of Chinese classics? So he moved to New York, where he had many friends, and rented an apartment where he could return to scholarship in peace and quiet.

Overshadowing all this was the possibility that, after the surrender of Japan, the civil war in China would resume again. In the summer of 1945, Hu attended the founding conference of the United Nations in San Francisco, as a member of the Chinese delegation. He met Dong Biwu, delegate of the Chinese Communist Party. and had a long discussion with him.

On August 24, he wrote a letter to Mao Tsetung who, as a clerk at Beijing University, had once written to him. "Chinese Communist leaders like yourself should try to forget the past and look toward the future, resolutely foregoing the use of military force and setting as your goal the creation of a second political party in China that does not depend on military might. Such a decision would bring to an end the protracted domestic conflicts of the last 18 years and forestall a civil war in which all your hard work of the past two decades might come to nought." Mao paid no heed to his advice.

As well as all the diplomatic work he did to try to salvage the political and security crises in China, Hu was also acting head of the Chinese delegation to the first U.N. Educational, Scientific and Cultural Organisation (UNESCO) conference in London in November 1945.

Family

Hu's elder son Zu-wang arrived in San Francisco on August 18, 1939. He brought with him cases full of his father's texts, including unfinished manuscripts. He studied engineering at Cornell University, at a cost of US$1,200 a year, graduating in May 1942 and going to work for the Studebaker Motor Company.

In April 1941, Hu asked his second son Si-du to come from Shanghai and join him in Washington. In China, Si-du had not been a good student, perhaps because his father was absent much of the time. He loved Beijing Opera and was overweight from eating too much; his mother found it hard to get him out of bed in the morning. Hu sent him to Haverford College, a private liberal arts college in Haverford, Pennsylvania; he was not a good student. He did not graduate from there nor from a second college where his father enrolled him.

Hu wrote to his wife to explain why he was inviting Si-du but not her to come. "My life here is not very pleasant ... You and I can live the family life of a college professor and not the family life of a diplomat."

He explained he was forced to attend social functions, which left him exhausted; she would suffer leading such a life. After Si-du left, Ms Jiang was on her own without her two sons. When Hu left in September 1937, she had expected her husband to be away for only several months. In July 1941, after the U.S. and Britain imposed economic sanctions against Japan, mail services were cut between Shanghai and the U.S. She did not hear from him until 1945. So she endured the eight terrible years of the Anti-Japanese War without her husband at her side.

In his life, Hu was a success at many things, but not, it seems, at fatherhood. In 1939, he wrote to his wife: "You and I have not done right by our sons. Looking back, I wish I could make amends, although I am afraid it's too late. We should change our attitude to them from now on. We should treat our sons as friends. They have grown up, and scolding won't do them any good."

Academic life

Hu rented an apartment at 104 East 81st Street in New York for scholarly work. This was in an expensive residential district. The apartment was arranged by Mrs Virginia Davis Hartman, the nurse who started to take care of him in the Presbyterian Hospital on December 6, 1938 after his heart attack. She was a widow when she first met him and four years his junior. After he left hospital, they continued to see each other. Their friendship deepened while he lived in New York from 1942 to 1946. They saw a lot of each other; she accompanied him and Zu-wang to plays and films. Their friendship lasted 20 years, until he settled in Taipei in 1958.

His friendship with Miss Williams continued in a more limited fashion, with the exchange of gifts and flowers on their respective birthdays. He had many other friends, Chinese and American.

In his diary of October 20 1942, he wrote that he wanted to work on *An Outline of the History of Chinese Philosophy*; he had published the first volume of it in 1919. The Rockefeller Foundation agreed to fund his research. He had access to the excellent academic libraries of New York, which included large Chinese collections. In November 1943, he switched his attention to research into *The Water Margin*.

In 1945, the Chinese government appointed him President of Beijing University, with Fu Si-nian holding the post for one year until Hu returned from the U.S.

Initially, Hu did not want to accept the post. He knew what an enormous challenge it would be to rebuild the institution after eight years of war. In a letter, Fu told him that the university needed US$50 million to US$100 million; where would this money come from?

In the end, Hu could not refuse. Many colleagues asked him to join the mission of resurrecting the university. While he had spent the eight years of war in the safety and comfort of Washington and New York, they had endured Japanese bombing, loss of family and friends, inflation and suffering of every kind. Now that peace had come, how could he refuse their requests?

After the end of the war in August 1945, BU was unable to return to its home campus because it needed repairs and because of the enormous logistical difficulties across China after eight devastating years of war. So it remained in Kunming for another year, holding its final graduation ceremony there on May 4, 1946.

On May 2 1946, Hu suffered a mild heart attack in New York. Mrs Hartman gave him pills and an injection to ease the pain. Hu took a boat from New York on June 5, 1946; he had a private cabin and 40 pieces of luggage. He had been in the US eight years and eight months. He was 55. In his diary on June 5, he wrote: "Goodbye, America! Goodbye, New York!". Did he expect never to see them again?

Sources for Chapter Seven

Japan-U.S. Trade and Rethinking the Point of No Return toward the Pearl Harbor, by Ryohei Nakagawa.

Chinese People's Political Consultative Congress Daily, 13/11/2014, article on Hu Shih's period in the U.S. from 1937-1942.

Wartime Ambassador: Hu Shih in Washington, October 1938 - September 1942, by Sarah Abbass, University of Western Sydney.

China's War with Japan 1937-45, the Struggle for Survival, by Rana Mitter, Allen Lane, 2013.

U.S. State Department, Office of the Historian: *Revised Draft of Proposed "Modus Vivendi" with Japan*, 24/11/1941.

Website of Beijing University

Revisiting the Course of Hu Shih's Life: A Re-evaluation of Hu Shih's Life and Thought, by Yu Ying-shi, Linking Books, Taipei, 2014.

Diary of Hu Shih

A Pragmatist and his Free Spirit, the Half-Century Romance of Hu Shih and Edith Clifford Williams, by Susan Chan Egan and Chou Chih-p'ing, Chinese University Press of Hong Kong, 2009.

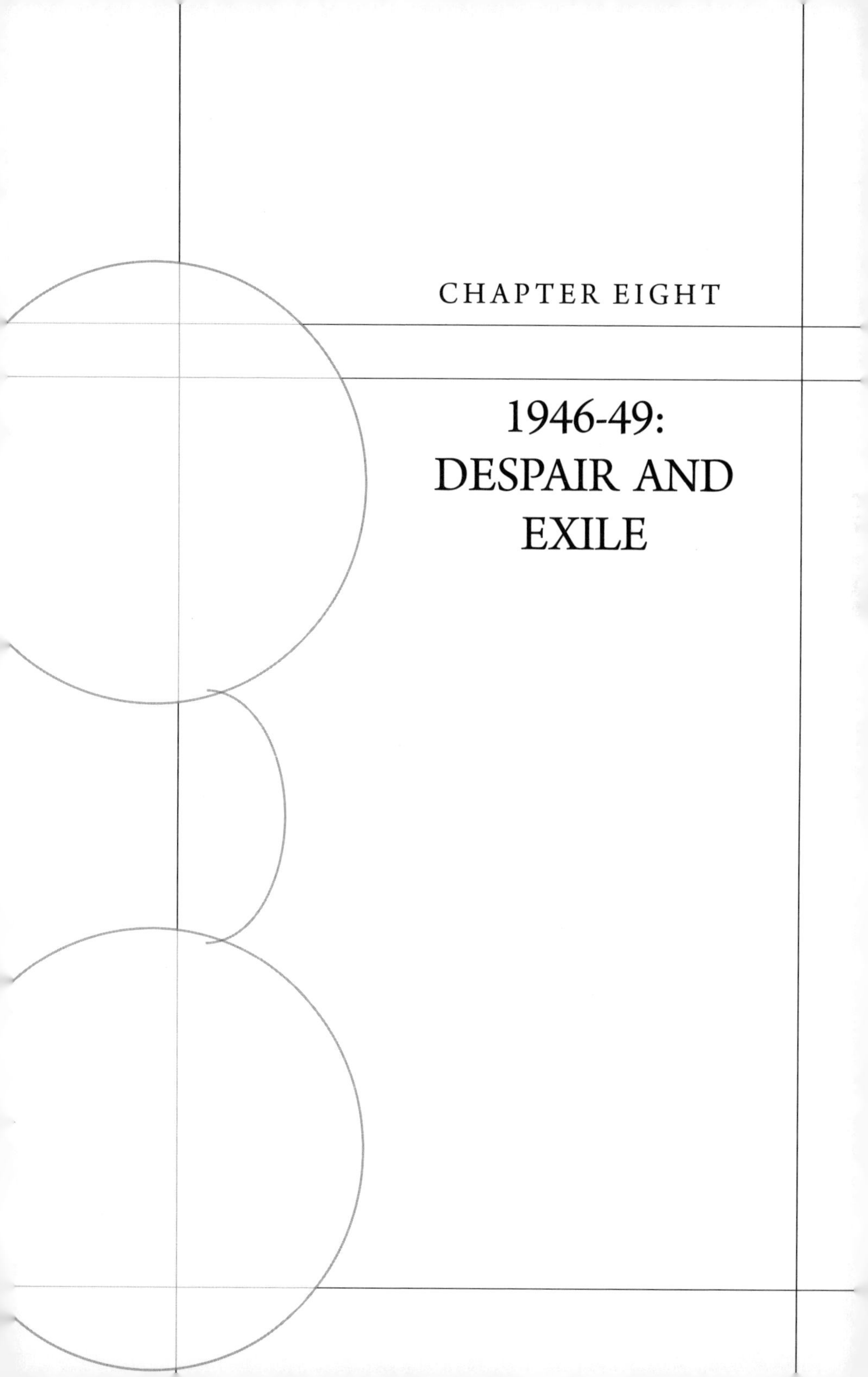

CHAPTER EIGHT

1946-49: DESPAIR AND EXILE

During the seven-week journey in his private cabin on the "President Taft" from New York to Shanghai, Hu had much to think about. He was about to become president of Beijing University (BU), the most famous college in China and with which he had been intimately connected for 30 years. It was the culmination of his intellectual career. But the challenges ahead loomed larger than the glories of the post.

The war had forced BU into exile for nine years in Kunming and seriously damaged the campus. Acting president Fu Si-nian told him the university needed US$50 million to US$100 million. Who could provide this money – the government or corporate or private donors in China and the United States? Hu needed to pay competitive salaries to attract the best professors and researchers. An even darker cloud was the civil war between the government and the Communist Party that seemed impossible to avoid.

In December 1945, U.S. President Harry Truman had sent to China General George Marshall, former head of the Joint Chiefs of Staff, to mediate between the two sides. He brought them together at a "political consultative conference" in Nanjing in January 1946. His efforts failed because of military clashes in many parts of China and the assassination of liberal intellectuals. Communist forces kidnapped and killed American servicemen stationed in China.

On August 10, Truman told President Chiang: "American faith in the peaceful and democratic aspirations of the Chinese people has not been destroyed by recent events, but has been shaken." A civil war seemed inevitable. Who would win this war and what would be BU's future under the government that won?

In his diary for July 24, 1946, Hu wrote that heavy rain fell as the ship approached Shanghai: "After the sky cleared, at 20:15, a bright evening glow rose over the sea, something rarely seen in life. After nine years of not seeing the motherland, the setting sun was clearly resplendent!"

"Resplendent" could mean the brightness of the sun – or the sight of artillery fire in an upcoming war. Fu came to Shanghai to meet him, as well as his son Zu-wang and a group of reporters. After meeting city officials, Hu finally saw his wife late in the evening, for the first time in nine years.

We have no record of the meeting. Miss Jiang had spent most of the war in Shanghai, first in the French Concession and then under Japanese occupation. She had not suffered physical shortages. In the spring of 1945, she returned to her ancestral home in Anhui. Fu and Hu then went to Beijing; Hu was welcomed by both the city government and the BU alumni association. He was no longer a famous professor but a national figure – a position with which he was not entirely comfortable.

The return of BU to his its historic home after an absence of nine years was a matter of great importance. The staff made intensive preparations and held a ceremony to mark the re-opening on National Day, October 10. To accommodate returning teachers and students, the university needed new buildings. It obtained the former residence of President Li Yuanhong in Dongcheng Hutong, and the old Council House.

The school was scattered over 40 locations in urban and suburban Beijing. The university added Medical, Agriculture and Engineering Schools to the original schools of Literature, Science and Law. This made a total of six schools, 23 departments and two specialised programmes. It also established graduate schools of liberal arts, sciences, law, and medicine. By mid-December 1946, it had 3,420 registered students, including 564 who had come from the NSAU in Kunming.

In the first semester of 1947, there were 58 graduate students and, in the second semester, 45. Most students needed four years to obtain a bachelor's degree, except for those in engineering, agriculture and pharmacy (who needed five), dentistry (six) and medicine (seven).

To rebuild a major university after eight years of war is an enormous challenge in normal times. But these were anything but normal. One obstacle was rampant inflation. From a base of 100 in September 1945, the index of wholesale prices in Shanghai reached 1,180 in July, the month of Hu's return, and 1,900 by January 1947. This was a disaster for those on a fixed salary, like the faculty and staff of BU.

The Communist Party skilfully exploited the discontent. It infiltrated trade unions and organised strikes, 1,716 in Shanghai alone in 1946. Party members were active on the campus of universities, including BU, organising marches and protests against the government. Hu faced the same student activism as he had during his two previous stints at BU. But this time it was better organised and more widely supported, including by members of the faculty.

Across the country, the civil war intensified. In July 1946, government forces began a broad assault on the People's Liberation Army in Manchuria. In districts the Communist Party controlled in North China, it began to confiscate rural land and deal violent punishment to "class enemies".

In early January 1947, George Marshall left China; he declared that his mission to bring the two sides together had failed and allAmerican mediating groups in China had been disbanded. This meant a war to the finish, with thousands of dead and a single victor. What would be the

Fu Sinian (left) welcomes Hu Shih on his return to Beijing University

place of BU, and Hu himself, in the new order? Pictures of him at BU during this period show him thin and exhausted.

Throughout his life, Hu strove to stay out of politics. He never joined a political party. Like other liberal intellectuals, he saw his role as an independent able to comment on and criticise both the government and those opposing it. That was the intellectuals' place in society.

From the 1920s, Hu had opposed "isms", including Communism, as radical and simplistic. In 1946, in New York, he read *I Chose Freedom,* by Victor Kravchenko, a Ukrainian-born member of the Soviet Communist Party who defected to the U.S. in 1944. His book, published in 1946 in New York, became a best-seller in the U.S. and Europe. It revealed collectivisation, the gulag prison system and use of penal labour in the

Soviet Union; all three were little known at that time in the West.

In his diary for April 24, 1946, Hu wrote that he was very shaken by the book. "Kravchenko is appealing for public opinion to protect him. This book is his autobiography. It describes the internal oppression of the Soviet Union. Quite shattered. The Soviet Union was the country on which Mao Tsetung wanted to model his new republic.

In 1947, Hu received a letter from Fu Si-nian, a close friend; the two men held similar views: "We must create a force different to that of the Chinese Communist Party, from learning to the outlook on life and from ideals to practice; they are not at all the same. If they take power, China will follow the fate of the Soviet Union. To prevent this, we want the government not to fall but to improve itself."

The letter shows how little, if any, space existed for independent thought and comment between the two sides that were fighting a battle to the death. You are with me or against me – there was nothing in the middle.

Hu's interaction with the government mainly involved education. In November 1946, he took part in a national conference in Nanjing to write the new constitution of the Republic of China; the Communists did not participate.

Hu gave a speech at the conference. He and other education leaders proposed that a specific clause in the constitution cover expenditure for this sector. He worked hard to secure funding for BU from the government and from the China Foundation for the Promotion of Education and Culture (CF) which we mentioned in Chapter Five. Its great advantage was that its funds were denominated in U.S. dollars, unlike the Chinese currency that was depreciating every week.

In March 1947, Hu proposed to the eight board members an agreement similar to that which CF and BU had had in the 1930s. He asked for a loan of US$300,000, to be disbursed in U.S. dollars over two years with an interest rate of five per cent. After the loan had been paid, BU would repay the money over 15 years, with the Ministry of Education as guarantor. The money would be used to buy equipment for the university The CF board considered the request. At a meeting on December 13, it decided to disburse US$250,000 to four universities. BU would receive US$100,000, to be spent on the physics faculty. Three others would each receive US$50,000 – Zhejiang University, Wuhan University and National Central University in Nanjing, Jiangsu province.

BU aimed to use the money to create a Modern Physics Centre. But, in February 1949, Hu discovered that, for various reasons, the money had not been used; so he returned the US$100,000 back to the CF.

In September 1947, he proposed to the government a 10-year plan to concentrate resources to create five top-rate universities – BU, Tsinghua University, Wuhan University, Zhejiang University and National Central University.

While many welcomed the project, the government did not have the funds because inflation was rising every week. In his diary entry for September 23, Hu expressed his disappointment at the reaction of his fellow teachers at BU. About 100 gathered for a meeting that day; he chaired the session for two and a half hours. "After I went home, my heart was full of sadness. This kind of university president is not worth doing! What everyone talked and thought about was what to eat! What Mr Da said made me even more angry. He said: 'What we worry about today is our life tomorrow. Who has time to think of a 10- or 20-year plan? Ten to twenty years later, everyone here will be dead.' "

Hu's disappointment was understandable. The government had brought him back to turn BU into a world-class university like those in the U.S. where he had often spoken. He had the vision and personal networks to raise money and attract leading professors. But these were abnormal times. The words of Professor Da more accurately reflected the daily fears and anxieties of his colleagues than those of their president.

Eleven years later, Hu would propose a similar plan to the government in Taiwan. We will speak of that in the next chapter.

So, Hu's tenure did not turn out as he had hoped. In this intense atmosphere of civil war and political polarisation, how could students and faculty concentrate on learning? Hu had to devote much of his time to raising money; he had to handle student protests and demonstrations and try to keep students out of jail. Photographs of this period show him gaunt and anxious, far from his normal optimistic self. On one occasion, he submitted his resignation to Education Minister Chu Chia-hua. But the Minister immediately sent him a telegram refusing the request.

Presidential candidate?

From January 1947, President Chiang made several attempts to persuade Hu to join his government. He offered him the post of director of the Examination Yuan, which oversaw the exam system and promotion of civil servants. He also invited him to join the National Government Commission, the highest policy-making body.

In his diary of February 22, Hu wrote that the government wanted to hire one or two people who spoke independently and be useful during such a critical time. "But I absolutely am not someone who does not take care of himself. I do not wish to give up my freedom to come and go as I wish

...You cannot run BU and be a member of the Commission at the same time. I asked Mr Chiang to let me stay on here and do something useful for the country."

In March, Hu went to Nanjing for meetings of the CF and Academia Sinica. Over six days, Chiang twice invited him for meetings and again asked him to join the government. He said that Hu need not take the post at the Examination Yuan but simply join the National Government Commission, which held two meetings a month. A member was not an official; Hu need not attend every meeting. He again refused. The president was polite and showed him personally to the door.

One reason for Chiang's insistence was Hu's high standing in the U.S. Political and public opinion in that country was turning against the Nationalists. Hu's presence in the government, even without a specific role, would improve its image in the U.S. Later in the year, Chiang made him further offers, to be a candidate in the Presidential election of April 1948 and Prime Minister. He declined.

On December 16, Chiang invited Hu to a meal; when he arrived, he found that he was the only guest present. Chiang pressed Hu to return to Washington as ambassador. Hu did not answer, only saying that he would consider it. In his diary of December 12, he had written his reply: "I am old, 10 years since I last went (as ambassador). It was not like before. I could go as one member of a team for peace talks with Japan in Washington, but not as ambassador."

By the start of 1948, Hu had given up hope of a negotiated peace between the two sides. He was increasingly dispirited. One of the biggest civil wars in human history was raging across the country; thousands of people were being killed, injured and displaced.

By late 1947, Communist armies and guerrillas controlled most of the countryside of north China. The cities of Shenyang and Changchun in Manchuria were surrounded by the PLA and could only be supplied by air. American military advisors implored Chiang to evacuate his 200,000 well-trained troops from Shenyang, to defend north China, but he refused. In September and October 1948, Lin Biao led campaigns that led to the fall of these two cities, with the destruction, surrender and desertion of 400,000 of Chiang's best soldiers. This opened the way for the PLA to march on Beijing.

"No Bread and No Liberty"

December 17, 1948 marked the 50th anniversary of Beijing University. With his colleagues, Hu was busily preparing events to mark this important anniversary. In November, rumours were circulating in Beijing that, as in 1937, BU would move to the south. Hu denied these; focused on the celebrations, he did not want to leave. On November 24, the professors' council passed a resolution opposing relocation. But it was clear that Hu would soon have to make a choice.

Fresh from its victories in Manchuria, the PLA on November 29 launched the Pingjin campaign, with the aim of capturing Beijing, then called Beiping, and Tianjin, the two most important cities in North China. The Nationalist commander in the region, General Fu Zuoyi , had more than 500,000 troops under his command. But he was unable to resist the PLA advance. Since the Communists controlled most of the countryside, everyone realised that the PLA would soon surround Beijing and General Fu's army inside it.

In his diary for December 4, Hu wrote that he attended a banquet that evening in Beijing to welcome a scholar who had just returned from the

During the civil war, Hu Shih addresses students at Beijing University.

U.S. In attendance were leading members of the university. In his closing speech, Hu said that, after the 50th anniversary events, he no longer wished to be president of BU. "I would like to go where the government is and do something useful. When I am not president, I certainly will not do *An Outline of the History of Chinese Philosophy* or *The Water Margin*. As to what I can do, I myself do not know."

Behind the jokes and polite words, everyone at the banquet knew that their world was about to change profoundly. Among the BU faculty, like the Chinese intelligentsia as a whole, there was a wide divergence of views. Some were underground members of the Communist Party working for the overthrow of the government, some were neutral. Others were critical of President Chiang and his Nationalist party but were uneasy about their own future; in the new order, would they retain their job and social status? Others supported the government and regarded a Communist victory as a tragedy for China.

Hu belonged to the third group. Because of his position and national status, the Communist Party very much wanted him to remain in Beijing and endorse its victory. It sent a student named Wu Han to appeal to him. In reply, Hu told him: "I believe nothing the CCP says. In the Soviet Union, there is bread but no freedom. In the US, there is bread and freedom. After the Communists come, there will be neither."

With Hu stating his views so clearly, Wu left. But the Party did not give up. One of its radio stations announced that, in future, Hu would be both President of Beijing University and also Director of the Beijing Library.

Unmoved, Hu did not change his mind. He knew that, because of his status, he was likely to be offered the chance to leave by air – an option not available to the majority of his colleagues. If they tried to leave Beijing by land, they would have to take their chances in crossing both government and PLA lines; they could be arrested by one side or the other. Or they could find themselves in the middle of a battle for control of Beijing.

On the morning of December 14, before Hu had left the house, he received a call from a friend in Nanjing advising him to leave Beijing and saying that an aeroplane would come to evacuate him. No plane arrived that morning. At 13:30, he received another call, telling him to meet others at military headquarters at the Palace of Fragrant Hills at 15:00. But, because the roads were blocked, they were unable to reach the airport.

In his diary for December 15, he wrote that, at 23:00 the previous evening, he had received a call from General Fu Zuoyi himself. He said that the President had called to say that Hu must fly to the south and that a plane would arrive the next morning (December16) at 08:00. Hu expressed his apology to the General for not remaining in Beijing with him; he accepted the apology.

The next morning, carrying only a few suitcases, Hu and his wife went to the military headquarters; they were told to wait for news there. In the early afternoon, they set out and reached Nanyuan Airport after 15:00. There were two planes there, each with places for 25 people. With his wife, Hu boarded one of the planes, along with other leading intellectuals. The planes flew directly to Nanjing, arriving at 18:30. He wrote in his diary: "Many friends came to greet us. Our son Si-du remained in Beijing and did not come with us."

Before his departure, he left a message to his senior colleagues saying he had made no preparations to depart and that he was leaving everything in their hands. "Although I am far away, I will certainly not forget Beijing University."

Hu would never see Beijing again, nor the university to which he had dedicated 31 years of his life.

Si-du, who turned 27 the next day, did not leave with his parents. It was an extraordinary decision. He surely knew that he was not an ordinary citizen but the son of one of the most famous people in China; that was his identity, like it or not. His parents tried their best to persuade him to come with them. In response, he said that, since he had done nothing to harm the Communist Party, they would not do anything to him. His parents left him many items of expensive clothing and gold and silver ornaments, which he could use for his wedding. His turned out to be a fatal decision.

Their elder son Zu-wang had left Beijing earlier. In the sudden rush to leave, Hu Shih had to leave behind more than 100 crates of books at his home in Beijing. They included some which belonged to his father and others he had carefully chosen and annotated. He was never able to recover them; it was a terrible blow for a scholar to lose the collection of a lifetime. Hu was one of a minority of the 81 members of Academia Sinica to leave the mainland. Nine went to Taiwan and 13 went overseas or were already there; 59 remained in the mainland or returned shortly after the PRC was set up.

In Nanjing, the capital of the government, Hu was safe and secure. But he felt uncomfortable – his high status had enabled him to leave Beijing, but most of his colleagues and friends did not have this option. As he wrote in his diary on January 1, 1949, he felt like a deserter and refugee. A friend who met him in Nanjing recorded this conversation with him. I am sitting in this room and with this coal, said Hu. All is being paid for by the state. Am I such a person that the state has to look after? His friend replied that this was temporary. If Hu could go abroad and represent the government, he "could still save the nation". Hu replied: "This kind of country, this kind of government, why should I want to raise my head to speak for them to foreigners!" Many in the government wanted him to take this role. At a dinner on January 8, Chiang asked him to go to the U.S. Hu repeated that he did not wish to be ambassador nor have any mission from the government. Chiang said that he simply wanted him to go and see what was happening.

On January 16, after 29 hours of fighting, the PLA captured Tianjin; the government armies lost 130,000 men killed or captured. With Beijing surrounded, General Fu Zuoyi decided on January 21 to negotiate a peaceful surrender. A week later, more than 250,000 of his troops started to leave the city. The fall of China's ancient – and future – capital was a milestone in the civil war. The way was open for the PLA to move south. The Communist Party published a list of "war criminals", which included all the senior members of the government and many of Hu's friends – but not Hu himself.

On January 15, Hu moved to Shanghai, where he stayed for more than two months in the apartment of a banking friend. The postal service between Shanghai and Beijing was still open; he received many messages from the heads of different departments at BU. He arranged for his wife to take a boat to the safety of Taiwan.

On the morning of April 6, he boarded the "President Cleveland" at the port of Shanghai; it left at 11:00. It was not easy to obtain a visa for the U.S. Universities there had to send him invitations to give a good reason for the visa. Washington was giving up on the Nationalist government. It was the sixth time Hu had left the mainland; he would never return.

Sources for Chapter Eight

Beijing University website

Hu Shi and His Friends, 1904-1948, Commercial Press Hong Kong, 1999.

Revisiting the Course of Hu Shih's Life: A Re-evaluation of Hu Shih's Life and Thought, Yu Ying-shi, Linking Books, Taipei, 2014.

Diary of Hu Shih

The Search for Modern China, Jonathan Spence, W.W. Norton & Company, 1990.

Selected Works of Hu Shih, Wang Ching-tian, National Library Publishing Company, New Taipei City, 2020.

A Pragmatist and his Free Spirit, the Half-Century Romance of Hu Shih and Edith Clifford Williams, Susan Chan Egan and Chou Chih-p'ing, Chinese University Press of Hong Kong, 2009.

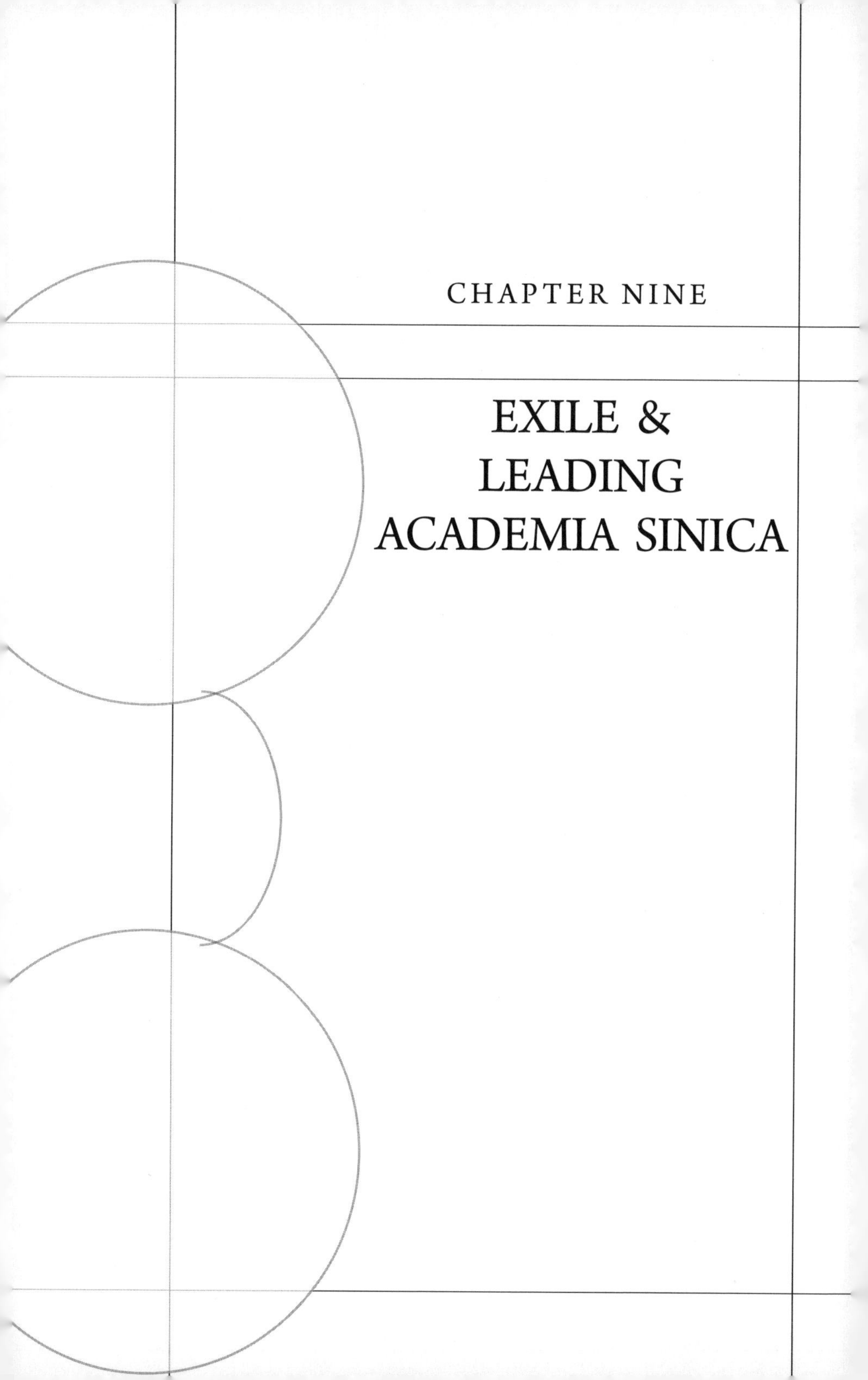

CHAPTER NINE

EXILE & LEADING ACADEMIA SINICA

When Hu left Shanghai, he knew that he might never return to the mainland. He reached the safety of the United States, where he would stay for nine years. His preference was to settle in Taiwan and work with his friends and colleagues there. But his wife did not want to live there and asked to join him in the U.S.; he respected her wishes. He finally moved to Taiwan in the spring of 1958, as President of Academia Sinica, a post he held until his death. His physical health declined – but his spirit remained as strong as ever, until the very last moment.

In April 1949, Hu arrived in San Francisco. He went to live in New York, in 104 East 81st Street, for the second time in 10 years, with Mrs Virginia Hartman.

The early months in the U.S. were difficult. In meetings in Washington and New York, he found that the attitude to the Nationalist government had completely changed. Everyone knew that it had lost the civil war and was retreating to Taiwan. No official or civil organisation would organise a talk for Hu to argue its case – as everyone had rushed to do in 1937.

In the summer of 1949, U.S. Secretary of State Dean Acheson wrote to President Harry Truman that the Nationalist armies were responsible for their own defeat. "A regime without faith in itself and an army without morale cannot survive the test of battle."

On October 1, Mao Tsetung proclaimed the establishment of the People's Republic of China from the Gate of Heavenly Peace overlooking Tiananmen Square in Beijing. On January 5, 1950, President Truman issued a statement that the U.S. would not use its military forces to interfere in the civil war in China. He defined the new American "defensive perimeter" to include Japan, Okinawa and the Philippines – but exclude Taiwan and South Korea. Washington was waiting for the PLA to invade and capture Taiwan.

So the context was different to that Hu faced when he arrived in the U.S. in 1937. Then he had a clear mission – to win over government and public opinion to the side of China in its war with Japan. This time he was a private citizen who had left his homeland because he did not want to live under a new regime.

Friends urged him to take a teaching post in the U.S. and bring his family over. But his preference was to settle in Taiwan and work there, as a lecturer and writer. He felt a strong loyalty to his friends and colleagues who had found refuge on the island; they wanted to build there the kind of state they had been unable to in the mainland. Hu retained the dream of a free, democratic China; Taiwan was the only place where this might be possible.

He had access to and the respect of President Chiang. As one step to advance his ideals, he became publisher of *Free China Journal* (FCJ) a fortnightly magazine established in 1949 in Taiwan to promote freedom and democracy. It would be a platform for different voices, especially those critical of the government. Its mission was similar to that of magazines which Hu had edited and wrote for in the mainland. It soon ran afoul of the Taiwan Garrison Command (TGC), the body Chiang established to implement martial law on the island. In 1951, the TGC forced the magazine to publish an apology after it criticised the TGC for imposing what it considered improper banking controls. In protest, Hu resigned as publisher.

In October 1949, Hu received a letter from Madame Jiang in Bangkok; she had gone there to attend the marriage of their son Zu-wang to Zeng Shuzhao, a graduate of Jinling Women's University, on October

1. In the letter, she said she did not wish to go back to Taipei. Hu was disappointed; this left him no choice but to remain in the U.S. His wife had lived throughout the Anti-Japanese war without her husband; he could not leave her on her own a second time.

So, he began to look for a job in the U.S. to enable his wife to obtain a visa. Between October and November 1949, he had two episodes of cardiac spasm. He was fortunate to have Mrs Hartman, an experienced nurse, to look after him. In May 1950, he accepted a two-year contract as curator of the Gest Oriental Library at Princeton University. The salary was US$5,200 a year.

The library contains a great diversity of material in Chinese, Japanese and Korean; it is invaluable for research done in the university's Department of East Asian Studies.

At Princeton, Hu organised an exhibition on "Eleven Centuries of Chinese Printing" and wrote a scholarly paper on the history of book collecting.

His new post enabled Miss Jiang to obtain the visa. In June 1950, she arrived in the U.S. She joined him in the apartment in New York; he commuted by train to Princeton, in New Jersey. Fortunately, he did not have to go to the library every day; he also gave occasional lectures at Princeton. Before his wife arrived, Mrs Hartman moved out of the apartment; thereafter she and Hu kept in touch – discreetly.

Then, on June 25, 1950, everything changed. The North Korean army launched a massive invasion of the South and drove the government to a last stand in the southeast port city of Pusan. Alarmed, President Truman ordered U.S. troops in Japan to intervene to help South Korea. The United

Nations Security Council – with the Soviet Union absent – condemned the attack and urged U.N. members to help; 15 member nations sent soldiers to help the U.S. side.

Truman also ordered the U.S. Seventh Fleet to patrol the Taiwan Strait; this made it impossible for the PLA to invade Taiwan. In October, Chinese troops entered the Korean War on the Northern side. In total, their number would rise to more than 700,000. The U.S. and Chinese interventions in the Korean War saved the government of Chiang Kai-shek. Taiwan had become part of the U.S. defence perimeter in East Asia. The interventions were like a second attack on Pearl Harbour – but this time Hu played no part in causing them.

"Enemy of the People"

On September 22, 1950, the pro-Communist Hong Kong newspaper *Ta Kung Pao* published an article by Hu Si-du criticising his father. He wrote that Hu had become an imperialist tool and that he must draw a line between himself and his father and take the side of the peasants, the labourers and the masses. "Until he returns to the embrace of the people, he will be the people's enemy and my own enemy," he wrote.

It was the start of a ferocious, nationwide campaign against Hu Shih that lasted for several years. Reading the style and choice of words in the article, Hu concluded that his son was not the author but had been forced to sign his name. He knew well the political context; his son was not free to express his own opinions. "In China," Hu told an American magazine, "people do not have the freedom to express their own opinions nor even to remain silent."

The denunciation was widely reported by the U.S. media, including *The*

New York Times. In October, Hu's wife received a letter from Si-du saying that he was fine and his parents should not worry about him. He said that Father should take care of his health and not be too socially active. His parents took the letter as a further sign that the article in *Ta Kung Pao* had been written under duress.

In 1953, a nationwide campaign against Hu began in earnest; it denounced his philosophy, political thought, theory of history and of literature and literary history. More than 180 articles by over 160 authors were published, a total of five million characters. They were published in 1955 in eight volumes, under the title "Criticism and Repudiation of Thoughts of Hu Shih".

The Chinese Academy of Sciences and the China Writers Association set up nine sub-committees to lead the campaign. They called Hu the leading "spokesman for the comprador bourgeoisie" and said he was supported by "American imperialism". It accused him of advocating total westernisation and being the designated hack writer of Chiang Kai-shek. As we have seen, this was incorrect; for 20 years, Hu had had frequent conflicts and disagreements with Chiang, public and private.

Former friends of Hu in the mainland denounced him in public. Knowing how the system worked, Hu took the campaign with a light heart. He joked that it meant everyone in the mainland had to read his works: "While I die, my ideas will live on."

Why did the government attack Hu? One reason was that, since the 1920s, he had been a constant critic of Communism, while many of his friends and colleagues had joined the Party. Another reason was that, despite the Party's appeals in 1948, he had left the mainland and sided with the Nationalists. A third was his wide influence among the public, which the government wanted to change; the mainland term for his writings was "poison". A fourth was that, in the summer of 1950, he had written an article published in *Foreign Affairs* magazine in the U.S. entitled "China in Stalin's Grand Strategy", it said that Russian aid had been key to the Communist victory in China.

Si-du was sent to the Tangshan Railway Institute to work as a history teacher. Since he was tainted as the son of Hu Shih, people refused to mix with him; women would not go out with him. He reached his 30th birthday unmarried. In 1957, the Party launched the "anti-rightist movement". Hu believed he had said and written enough to draw a line between himself and his father. But the leaders of his college labelled him a "rightist". On September 21 1957, unable to bear these attacks and his own isolation, he hanged himself; he was just 36 years old. He left a suicide note and money to a second cousin. The mainland media did not report the news. In the U.S., Hu Shih and Zu-wang heard rumours of the suicide but could not confirm it. The government officially rehabilitated Si-du in 1980, 23 years after his death.

Sweep the floor, empty the ash-trays

The arrival of his wife made things more complicated for Hu. She did not speak English and had no interest in his work or intellectual life. She spent her time playing mahjong with Chinese friends; she also liked entertaining, cooking and reading martial arts novels.

Hu described himself as "a babysitter." With only a part-time helper in the apartment, Hu had to sweep the floors, wipe the tables and empty the ash-trays filled with cigarette butts left by his wife's mahjong partners. He would have preferred a quiet apartment with few visitors.

Hu Shih in his New York apartment in 1957

In 1952, to save money, Princeton decided not to renew his contract. That year Hu decided to stop teaching in the U.S. He explained his thinking in a letter to his friend Chao Yuan-ren on December 19, 1955. "In recent years, I have not done teaching, nor have I aggressively sought opportunities to give lectures. This is because of my desire to avoid suspicion. Many so-called 'Sinologists' fear my breaking into their circle (many scholars from

China had emigrated to the U.S. after 1949). In addition, among these people, there are those who are politically 'progressive' – not to my taste. So I feel a little sensitive. My policy is to be 'respected from afar', I do not want to have meals with them."

During those nine years in the U.S., his only full-time teaching post was Regents Professor in the philosophy department of the University of California at Berkeley for six months in the autumn of 1956.

He mainly lived off interest and dividends from investments and lecture fees. As he had done throughout his life, he treasured the company of his friends, including Mrs Hartman and Miss Williams.

In the summer of 1953, Hu and his wife spent a month in Ithaca, in the home of Miss Williams, at her invitation. It was the first meeting between the two women. Knowing that Madame Jiang was an accomplished cook, Miss Williams bought many Chinese ingredients and laid them out in the kitchen, so she could make the food that her husband liked. She chose as a departure gift a set of kitchen utensils and found, with difficulty, someone to carve the characters of her name Dongxiu in them. These showed how Miss Williams wanted to show respect to Miss Jiang and offset any jealousy she may have felt toward her husband's long-standing American lady friend.

Another remarkable episode was an agreement between Miss Williams and Mrs Hartman to share the considerable cost, US$500, of an air conditioner for Hu's apartment during the hot, sweltering New York summer. When he learnt about this generous offer, Hu declined it, saying that the summer would soon be over. It showed how he could inspire love among his lady friends that was co-operative and not competitive.

Settle in Taiwan

Hu maintained a strong interest in Taiwan. In November 1952, he accepted an invitation from National Taiwan University and Taiwan Normal College to give lectures. Hundreds came to greet him on arrival at the airport.

During the visit, President Chiang invited him for a meal. Their conversation lasted for two hours. Hu told Chiang that he (Chiang) needed 1,000 officials ready to criticise him – allowing freedom of speech would create an environment in which these good people could flourish.

Hu stayed for two months. He went again in the spring of 1954 and stayed for six weeks. His lectures attracted large audiences. He believed that not only would it be the best place for him to research and write; he also had a duty to promote freedom of speech and democratic values, which he could do better on the island than from abroad.

In September 1952, in a letter to President Chiang, he proposed a multi-party system, the abolition of the post of chairman of the Nationalist Party, freedom of speech and the right to criticise. In his diary, Chiang wrote that the world Hu lived in was different to the real world. "Chinese scholars are always like that, so they accomplish nothing in building the country."

Chiang believed that one reason he lost the mainland was that he had permitted too much freedom to civil society; this had allowed the Communist Party to organise and overthrow him. He would not make the same mistake in Taiwan.

In his diary on December 13, Chiang wrote that Hu had proposed Taiwan

On March 25, 1954, Hu Shih presents Chiang Kai-shek with the document of his appointment as President.

carry out democratic reforms and ally itself with the free world. During World War Two, the democratic side had won, Chiang wrote. "We were on the democratic side and made the greatest sacrifice. But, in the end, we were forced to flee our country" He called Hu an intellectual who did not understand well the difficulties of the real world.

In a letter to a friend on November 18, 1956, Hu wrote that he planned to settle in Taiwan, either in Taichung or Nangang, a suburb of Taipei that was home to Academia Sinica.

"The library at the Institute of History and Philology at the Academia Sinica is very well suited for my research, better than libraries overseas ...

I am old, nearly 'retirement' age. I still have some savings which may last me two or three years, but, in Taipei or Taichung, they will last more than ten."

The AS was founded in 1928 by Cai Yuan-pei as an academic institution independent of the government. In the one-party state established by Chiang, the members of the Academy had to struggle to maintain this independence; they also had to work on a meagre budget.

In October 1957, the post of AS president fell vacant. Under its charter, the members of AS proposed three candidates, from whom President Chiang chose one. Hu was his favoured choice. Accepting the post would, Hu believed, be the best way to fight for its independence and keep it out of the control of the Nationalist Party. It would also enable him to fulfill his ambition to return to China.

In a letter to a friend on January 4, 1958, he wrote: "I thought carefully, then asked the President to name this old brother to become (AS) president. This enables to me to recuperate and fully recover and return soon to the motherland to take up the post.

In April 1958, Hu moved to Taiwan and settled in the compound of Academia Sinica in Nangang. His wife chose to remain in New York until October 1961. He lived in a one-storey house the AS built for him; it had three bedrooms, red brick walls and a black tile roof. He had a servant and a chauffeur-driven car.

"He is really a mad man"

President Chiang came to the investiture ceremony at Academia Sinica on April 10, 1958. It turned out to be an event that the participants did not forget. In his speech, Chiang praised Hu for his academic achievements and moral character and said he should fulfill the mission of AS, "to restore China's culture".

Immediately, Hu stood up and said: "President, you are mistaken. The role of Academia Sinica is to promote freedom and democracy."

Everyone in the room turned pale; the atmosphere became very tense. In the authoritarian atmosphere of martial law, no-one defied the president, certainly not in a public forum. Only someone of Hu's status and self-confidence would dare to do so and get away with it.

In his acceptance speech, Hu said that he had always had a close relation with AS, where many of his friends had worked. He had been a researcher at its Institute of History and Philology, he said. "AS is the fruit of painstaking labour of many friends." He said that he had come to Taiwan against the advice of his doctors. "Many of my old friends have given the most precious years of their lives to AS. They have given everything. I have a responsibility to AS which I cannot avoid."

Chiang was so angry that he left immediately at the end of the ceremony. In his diary that evening, he vented his anger, describing Hu's behaviour as the second biggest insult of his life; the first was at a banquet in Wuhan in early 1927 at the hands of [Mikhail] Borodin (the Comintern advisor to the KMT sent by Vladimir Lenin). "The insult I received today at the investiture ceremony of Hu Shih at Academia Sinica in his reply. We can say it was a reproach when we try our best. I did not know this person was so arrogant and absurd. He really is a mad man. During the rest of Hu's tenure, Chiang never returned to AS.

"Respect the constitution"

Hu lobbied hard against Chiang's accepting a third term as president. In October 1956, the *Free China Journal* had published an article by Hu to mark the 70th birthday of President Chiang on October 31. He wrote that Chiang should observe the law and the Constitution, which barred him from a third term as president; his second term would expire in March 1960.

After settling in Taipei, Hu continued this effort. In his diary for November 11, 1958, he recorded a meeting with Chiang's secretary-general to whom he outlined his opinions, for him to relay to his boss. The Constitution set a time limit, which should not be lightly ignored, he said. "I hope that President Chiang will establish a model of a legal and peaceful transition to power ... I believe this would earn him the respect and admiration of his countrymen and the whole world."

Hu failed. On March 21 1960, the National Assembly, dominated by the Nationalist Party, elected Chiang for a third term. Chiang invoked "emergency provisions" to over-ride the constitution. He would serve a total of five terms and die in office in April 1975.

The *Free China Journal* (FCJ) also argued strongly against a third term. During the 1950s, its relationship with the government deteriorated. Things came to a head in 1960. Ahead of the National Assembly vote, it published several articles opposing Chiang's desire for a third term.

From May 4, its editor Lei Chen and fellow activists began to organise the Chinese Democratic Party. Hu was not a member but supported them. On September 4, the TGC arrested Lei and three others. A military court sentenced them to 10 years for "sedition".

The magazine was forced to close in September 1960; the CDP was never established. Hu was shaken by the sentence given to Lei. He was no student protestor. Born in 1897, he studied law and politics at the Kyoto Imperial University in Japan. He returned to China in 1926 and held important posts in the law and education. In 1947, he was elected as a representative to the National Assembly. In the 1950s, he had worked closely with Hu and other intellectuals in editing the *FCJ*.

In his diary entry for November 18, 1960, Hu recorded a meeting with Chiang that day. The president said he realised the long sentence had been badly received abroad; but he insisted that the government had no choice because Lei was connected with Communist spies. In response, Hu said that he hoped a civil case could try the case. "No-one in the world has confidence in the outcome of a military court."

He said that Lei's lawyer had only one and a half days to review the evidence; the trial for such a major case lasted only eight and half hours. The judgement was announced on October 8 and reported in the international media the next day, when Hu was in the U.S. He said that he had to hide himself in Princeton University during National Day, October 10, because he did not want to see anyone (and be questioned about the judgement). Hu felt guilty that he had been unable to help Lei; the case was another reminder of the fact that, while he had an access to the president enjoyed by few intellectuals in Taiwan, his influence over him was very limited. The case aged him.

Lei was released in 1970, but the government did not formally reverse the verdict in his case until September 2002, 23 years after his death in 1979. This is the judgement of the *Free China Journal* given in 2011 by Taiwan's Council for Cultural Affairs, which became the Ministry of Culture, in 2012. "The *Free China Journal* played an extremely important role as a promoter of democracy and constitutional government. It introduced many people to new ideas and, despite its closure, exerted a lasting influence on the future democracy movement."

"Study Broadly, Deeply and to a High Level"

From 1958-62, Wang Bin was a student at Taiwan Normal University. "In 1958, I went to the Sun Yat-sen Memorial Hall to hear a talk by Dr Hu. The hall was full up; students were standing at the back. Dr Hu spoke for more than one hour, in a soft voice and with a slight Anhui accent. He was smiling and had a friendly manner. There were jokes and he held our attention all the way through. He took questions at the end. President Chiang thought very highly of him."

After graduation, Wang returned to Hong Kong, where he taught Chinese for 30 years in secondary schools. At the back of his head is a hole left by a Japanese bomb that fell on his native village of Zhengcheng in Guangdong province in 1938. It killed his grandfather and grandmother, but he, eight months old, survived. He lived there until 1950, when his father, an officer in the Nationalist Army, came and took him to Hong Kong.

"I see Hu as the main architect of vernacular Chinese" Wang said. "That was a revolution in Chinese culture. His advice to us was to study broadly, deeply and to a high level. He also accurately analysed the Chinese character, calling us 'Mister More or Less'. Chinese are practical, taking for themselves but giving little back."

Science & Technology for the Future

One of Hu's long-term legacies to Taiwan was his part in establishing the National Council on Science Development on February 2, 1959. In September 1947, he had presented to the government a 10-year plan to achieve academic independence. It proposed that, instead of spending millions of U.S. dollars on sending students overseas, the government should concentrate its resources on building five world-class universities.

The civil war and rampant inflation meant that the government did not implement this plan in the mainland. Hu unveiled the plan again after moving to Taiwan. In his diary for March 26, 1958, he wrote that the country needed specialist people and research institutes to address the needs of industry, medicine, public health and defence; they should co-operate with global scholars and R & D centres to share responsibility for the work.

That year he invited experts and scholars to help him develop a long-term plan. His draft was "A Five-Year Plan for National Science Development and Training of Specialists" After intensive discussions and revisions, he presented the plan to the government at the end of May. The government response was positive. In August, it decided to allocate NT$40 million (for domestic spending) and US$500,000 (to be spent outside Taiwan) as a budget for the first year. The president gave his approval in January 1959 to setting up the council. Hu became its chairman, in addition to his role as president of Academia Sinica.

On November 6, 1961, Hu gave his last major speech, entitled "Social Change and Science at a Science Conference in Taipei, attended by participants from four countries. "In order to pave the way for the growth of science, in order to prepare ourselves to receive and welcome the modern civilization of science and technology, we Orientals may have to undergo some kind of intellectual change or revolution," he said.

"We should get rid of our deep-rooted prejudice that, while the West has undoubtedly excelled in its material and materialistic civilisation, we Orientals can take pride in our superior spiritual civilisation. We may have to get rid of this unjustifiable pride and learn to admit that there is very little spirituality in the civilisation of the East ... What spirituality is there in a civilization which tolerated so cruel and inhuman an institution as footbinding for women for over 1,000 years? An Oriental poet or philosopher sailing on a primitive sampan boat has no right to laugh at or belittle the material civilization of the men flying over his head in a modern jet airliner ... Such a reappraisal of the older civilisations of the East, and of the modern civilization of science and technology is an intellectual revolution necessary to prepare us Orientals for a sincere and wholehearted reception of modern science."

Hu's message was heard loud and clear across East Asia – China, Taiwan, South Korea and Japan have all since become powerhouses in science and technology.

In Taiwan, they have been ever since one of the most important pillars of the economy. An island with limited arable land and natural resources, it must rely on the skill and ingenuity of its people to drive the economy. In 1967, the council became the National Science Council and, in February 2014, the Ministry of Science and Technology. According to its website, the ministry's three main missions are promoting nationwide S&T development, supporting academic research, and developing science parks.

High-technology products have become one of the most important sectors of the economy. In 2020, Taiwan's exports reached a record US$345.28 billion, up 4.9 per cent from a year earlier, according to figures from the Ministry of Finance. Exports of electronic parts and components rose 20.5 per cent to US$135.6 billion, and those of information,

On March 9, 1960, Hu Shih and his grandson Hu Fu outside his home in Nangang, Taipei.

communication and audio/video devices increased 15.4 per cent to US$49.18 billion. These two categories – the science and technology sector – accounted for 54 per cent of total exports.

Family & Friends

In 1960, son Zu-wang moved to Washington to take up the post of economic counsellor at the Chinese embassy. He and his wife Margaret had one son, Hu Fu, who was born in 1955 with his foot deformed by polio. Like his father and grandfather, he went on to study at Cornell University.

In the autumn of 1960, Hu Shih visited the United States. Miss Williams, 75 and alone, had sold her home in Ithaca and decided to move to the Caribbean island of Barbados, which had a warm climate and a cheaper cost of living. Hu went to New York airport to see her off. It was to be their last meeting – perhaps both of them knew it; but they would continue their correspondence of 50 years. It was a poignant and emotional moment. A photograph of the meeting shows the two standing close together, each with a broad smile, the warmth of their love overcoming their age. Their friendship had been one of the treasures of their lives.

In her first letter to him from her hotel in Barbados on October 10, she wrote: "Your coming to see me off was a gift, rare, and, I fear, so costly no words can thank you. All this last month you have given unselfish consideration which no one could deserve, surely not I. This transportable picture of human kindness hangs in memory to bring happiness wherever I may be."

After Hu's death, she wrote to his son asking him to lay 50 small, fragrant white flowers on his tomb, to mark their 50 years of friendship.

"Teacher who started a new era"

In February 1961, Hu suffered a heart attack; he had to spend two months in hospital. In November, he had another heart attack; he spent a further six weeks in hospital. After he came out, CKS invited him and his wife to the Presidential Palace for lunch.

On February 24, 1962, the AS held its annual meeting. For Hu, it was one of the happiest days of the year, to be among his fellow scholars for whom he held such esteem. In the morning, he voted for seven new members. That afternoon he presided over a reception at the Cai Yuan-pei Memorial

The final meeting of Hu Shih and Edith Clifford Williams at New York airport in September 1960. She is leaving to retire in Barbados.

Building. In its report on March 1, this is the official *Taiwan Today* magazine described what happened:

"Hu was obviously tired but in good spirits. Joking with his guests, he repeatedly urged them to eat and drink heartily. Toward evening, while bidding farewell to departing members, he was stricken by a heart attack and collapsed. A half hour later, his spirit was gone. The half-smile around his lips attested that he had died as he would have wished: in the harness of scholarship. For days, the mortuary was thronged by those who come for a last look at China's great man of scholarship. From President Chiang Kai-shek to pedicab drivers, all had tears to shed for on intellect that had won universal respect and admiration. The line passing the bier was ceaseless by both day and night. They come not out of curiosity but to do honour to one who represented almost every outstanding value that Chinese culture has to offer. Tens of thousands lined the streets as the funeral cortege passed by. It was the largest such procession in Taiwan's history – a tribute to one of the few truly civilised men of our time."

He died with a smile on his face and surrounded by those who admired him. He felt no pain; his last words were "let's have some drinks". It was a blessed way to go.

The depth of public grief was remarkable. About 40,000 people visited the Elysium funeral home in central Taipei where Hu's body was laid. They included people of all classes and foreigners as well as Chinese.

President Chiang and Prime Minister Chen Cheng sent elegiac couplets. That of Prime Minister Chen read: "The teacher who started a new era, who mixed universal knowledge with the new learning and used (scientific) test to find the truth.

Madame Jiang and Zu-wang led the cortege down the streets of Taipei; mourners set up altars of fruit and incense along the way. Representatives from the U.S., Japan, South Korea, Vietnam, Thailand, Turkey and other countries attended the funeral. Over the coffin was draped the flag of Beijing University. While Hu had asked for cremation, his widow insisted on internment. A tomb was built on a hill facing AS; he was laid to rest there in October that year. Two pine trees stand over the grave. His home became the Hu Shih Memorial Hall. A resident of Nangang donated two hectares of land to build the Hu Shih Park. Members of AS and other famous scholars were later buried there.

Sources for Chapter Nine

The Search for Modern China, Jonathan Spence, W.W. Norton, 1990.

Wang Bin interview with author in Hong Kong, 24/2/2021.

Free China Journal, article published by Council for Cultural Affairs, Taiwan, 2011.

Selected Works of Hu Shih, Wang Ching-tian, National Library Publishing Company, New Taipei City, 2020.

A Pragmatist and his Free Spirit, the half-century romance of Hu Shih and Edith Clifford Williams, Susan Chan Egan and Chou Chih-p'ing, Chinese University Press of Hong Kong, 2009.

Revisiting the Course of Hu Shih's Life: A Re-evaluation of Hu Shih's Life and Thought, Yu Ying-shi, Linking Books, Taipei, 2014.

English Writings of Hu Shih, edited by Chou Chih-p'ing, Foreign Language Teaching and Research Press, Beijing, 2012.

Hu Shi Memorial Hall Archives

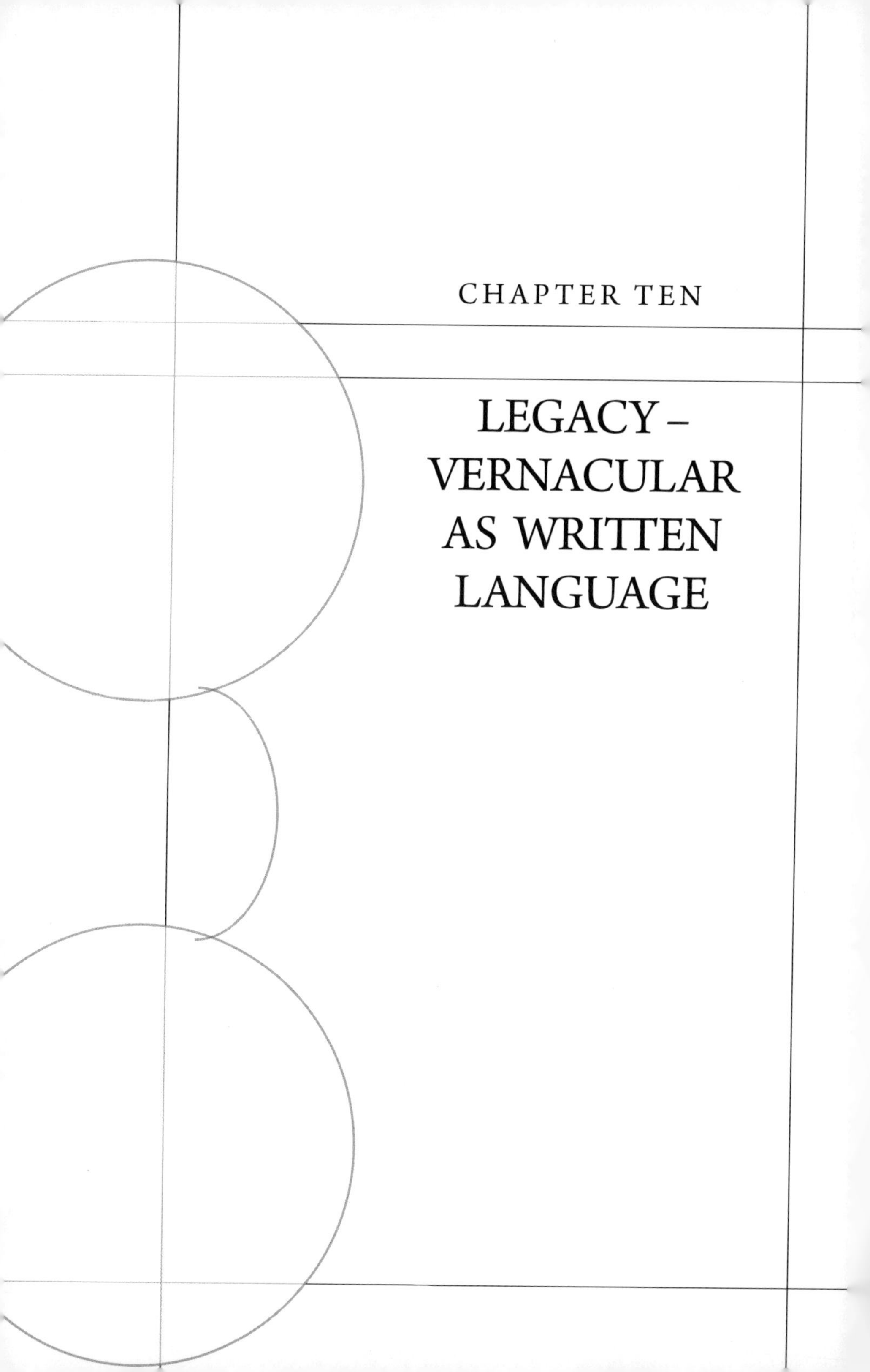

CHAPTER TEN

LEGACY – VERNACULAR AS WRITTEN LANGUAGE

Hu Shih died in April 1962. Few people in history have left such a written legacy. His output was prodigious. From his teenage years until his last day, he wrote poems, letters, articles, speeches and books as well as a diary.

He wrote 44 books, including three in English. He wrote in his study, bedroom, hotel rooms, passenger liners, railway trains, aeroplanes, teahouses and restaurants. Never reticent, he was happy to share his work with family, colleagues and a global network of friends and contacts. In addition, he wrote in two languages, Chinese and English.

Look at what the Memorial Hall in Taipei alone has published: 10 volumes of hand-written manuscripts, between 1966 and 1970; 37 volumes of his *Collected Works*, in 1986; and, in 1989, 18 volumes of his diaries.

In December 2018, Academia Sinica published *The Complete Works of Hu Shih*. Its Institute of Modern History spent six years collecting his work for this new compilation. Such a treasure house of material has inspired scholars in China, Taiwan, Hong Kong and overseas to research and write about him. He has become one of the most studied Chinese in history. Books by and about him are available in bookshops throughout the Chinese-speaking world.

Outside his writing, what was Hu's legacy to China? The greatest and most lasting one is the use of vernacular as the standard written language. As he predicted, it has provided a rich and diverse vein of creativity and literature. It has enabled tens of millions of people, unable to understand Classical Chinese, to read and write comfortably. In the 1950s, the PRC promulgated simplified characters, while Hong Kong, Taiwan and many overseas Chinese retained the traditional ones. But Hu's vernacular has remained the standard for both.

His legacy has remained strong in Taiwan, where he died in April 1962. The island was under martial law and the presidency of Chiang Kai-shek in a third term, which Hu had opposed. Martial law was finally lifted in July 1987, by his son and president Chiang Ching-kuo. It had lasted 38 years, at that time the longest period of martial law of any country in the world.

The government lifted the ban on opposition political parties and media. Since then, it has established a democratic system, with voting for president, members of parliament, mayors and other posts.

As of 2018, Taiwan had 220 political parties, including five important ones; 13 of them had representatives elected by the public. The media has become extremely diverse, with many different points of view represented; civil society is flourishing. Many Taiwan people dislike the current political order. They consider it chaotic and driven by party and personal conflicts; they would prefer a more centralised system that could reach consensus more easily and prioritise long-term economic planning, as it did during the Chiang era.

If Hu were alive today, he would surely be happy with the democracy and freedom Taiwan has achieved. Many factors have brought about these changes. Hu's writings and philosophy are certainly among them. Students of Chinese and history at universities in Taiwan learn of his writings and contribution to the New Literature Movement. Middle school students in Taiwan and Hong Kong read his essay "Mister Almost", a critique of the lazy character of many Chinese. This phrase has become part of the common language, with most of those who use it not knowing who created it.

In mainland China, we cannot speak of the same legacy. During the Maoist era, the government publicly denounced Hu's ideas and removed his works from bookshops, schools and universities. During the reform era, since 1980, he has been officially rehabilitated. But he is absent from the public discourse; readers of his work are limited to intellectuals and academics.

Memorial Hall

After Hu's death, Academia Sinica formed a "Hu Shih Memorial Hall Management Committee". They formally established the Memorial Hall on December 10 that year, using the house he had lived in. It holds Hu's papers after 1948. He shipped his books and documents from the U.S. to Taiwan in two periods. One was when he returned there in 1958. The second was when Madame Jiang was about to return to Taiwan to settle in September 1961. At his request, she shipped a further batch of books and documents.

On August 5, 1964, the management committee decided to allocate NT$200,000 from a fund donated by Cornelius Vander Starr to set up an exhibition room on the right side of the Hall; they formally opened it on the third anniversary of Hu's death in 1965.

Starr was, like Hu, a friend of Hsu Sing-loh, the banker who died in the air crash in China in August 1938. In 1919, Starr founded in Shanghai American Asiatic Underwriters, which later became the American International Group (AIG), one of the largest insurance companies in the world. In 1955, he set up the C.V. Starr Foundation which has since given millions of U.S. dollars to charities and causes around the world. It endowed East Asian libraries at Columbia, the University of Illinois at Urbana-Champaign and the University of California, Berkeley.

Academia Sinica and the Taipei City Government cooperated to build the Hu Shih Park near the site of his grave; work was completed in February 1974. In January 1998, the Memorial Hall became formally affiliated with the Institute of Modern History (of AS). The exhibition room has a collection of Hu's letters, manuscripts, copybooks and annotations. It also displays the Chinese changshan and western leather shoes, which were his usual outfit. He used different instruments for writing – a brush for poetry and a fountain pen for other compositions.

From "Enemy of the People" to "Great Contribution"

The story of Hu Shih in the mainland in 1949 mirrors its traumatic history of the last 70 years. In Chapter nine, we described the campaign against him in the 1950s. The attacks continued into the 1970s and the death of Mao Tsetung in 1976. Then things changed dramatically. In an article published in 2013, a mainland academic named Dong Ligong described the changing face of Hu since 1949. "After 1978, as Chinese society gradually returned to normal, research into Hu Shih was no longer entirely a taboo. It became possible to conduct an objective and fair evaluation of him."

Dictionaries described him in a more accurate and less ideological way. The 1989 edition of Ci Hai (Sea of Words Dictionary) called Hu a famous person in the New Culture Movement (of the 1920s) and that his research method of "brave supposition and careful search for the truth" was very influential. Its edition of 2009 praised his advice to "research more problems and speak less of 'isms'". And it also praised the promotion of "a government of good people" and constitutional government, ideas espoused by the magazine *Hard Work Weekly*, which Hu founded in 1922.

On April 22, 2015, the *Guangming Daily*, one of China's main newspapers for intellectuals, published an article by Zheng Dahua, of the Institute of Modern History at the Chinese Academy of Social Sciences, entitled "An Objective Evaluation of Hu Shih's Status in Chinese Academic History". It said Hu had a very important status in the history of modern Chinese academic history, with many firsts.

"We should fully confirm his academic achievements and contributions," Zheng said. "But confirm to what level? We must seek truths from facts ... In 1924, Hu gave himself a reasonably objective self-evaluation, in writing to Zhang Shi-zhao, a contemporary: 'Gong Sheng says that I have started a new era but am not a teacher – this makes me the happiest. We are both opening this era. I hope we remain friends and do not denounce each other.' This evaluation is quite proper."

Zheng called Hu a pioneer in many fields – the introduction of the vernacular language, translation of many important European novels into Chinese and his own plays, such as *The Greatest Event in Life* (on arranged marriages).

"But Hu had his limits. While criticising the traditional rules of poetry, he did not establish new ones." In February 1919, the Commercial Press in Shanghai published the first volume of his "An Outline of the History of Chinese Philosophy". It was so popular that, within three years, it went through seven editions. "This gave him an important status in the history of China's modern scholarship," Zheng said. "The reason for this popularity was that he was the first to use Western scholarly concepts to study Chinese philosophy. But his influence declined after the May Fourth and New Culture Movements and especially in the 1930s. He did not write the second or third volume; he had covered only a half or third of the history of Chinese philosophy. He was replaced by Feng You-lan, who wrote a two-volume history up to the Qing dynasty ... When

South Korean President Park Geun-hye was going through her darkest moments, the book she read was not that of Hu Shih but *The History of Chinese Philosophy* of Feng You-lan. This is one evidence (of Feng's importance)."

Born in 1895, Feng was a student at Beijing University and, like Hu, earned a Ph.D. at Columbia University and studied under Professor John Dewey. In 1934, Feng published *The History of Chinese Philosophy,* in two volumes; it became the standard work in the field.

Zheng said that another important contribution made by Hu was textual criticism of novels. "Between May 1917 and February 1962, he wrote more than 450,000 characters of textual criticisms of Chinese classical novels, including *Outlaws of the Marsh, Romance of the Three Kingdoms* and *Journey to the West.* His biggest influence was the textual criticism of *The Dream of the Red Chamber,* which he researched for more than 30 years. Hu Shih's academic achievements and contributions were very great but, like any historical figure, he had his limitations." Neither Dong nor Zheng mentioned Hu's 40-year opposition to Communism nor his extensive writing on freedom and democracy; these topics remain taboo.

Other people, including Hu's friends and supporters, have expressed a similar view – Hu was a person of such diverse interests and activities that he spread his energy too widely. This made it impossible to concentrate on a single subject, as many academics do. In Chapter Six, Professor Wen Yuan-ning called him "Doctor Half-Finished". Not only did Hu research many different subjects; he was extremely sociable and loved to meet and entertain friends. In addition, his working life from 1917 to 1962 spanned a period of extraordinary turbulence in China. It included the anti-Japanese and civil wars and intense student activism – the opposite of a comfortable professorship in a university in a small town in North

America.

When he left Beijing in December 1948, he was in a hurry and had not prepared; so he left most of his papers behind. Some were lost or destroyed during the campaign against him and the Cultural Revolution of 1966-76. After 1980, it became possible for mainland institutions to publish his work. In 1994, the Chinese Academy of Social Sciences in Beijing published 42 volumes of his papers, including 5,400 letters which others had written to him. Some date back to 1908. In 2002, the Beijing University Library brought out an elegant volume of facsimile reproductions of Hu Shi's hitherto unknown correspondence, as well as his diary from his teenage years in Shanghai.

But, outside these written works, Hu is scarcely honoured in the mainland. The monument to the May Fourth Movement at Beijing University does not mention him. Nationwide, there are only two statues of him. His hometown of Jixi, in Anhui, has opened his former residence as a tourist attraction. It is a modest Huizhou-style building, with two storeys and blue tiles.

In October 2020, an auction in Beijing raised nearly 140 million yuan (US$21.9 million) from the sale of 18 manuscripts of Hu's diary between 1912 and 1918; he was in the U.S. between 1910 and 1917. The diary was a Cornell University notebook, with the entries written in pen in English and Chinese in vertical columns. The handwriting is very dense; some is illegible. The survival of the manuscripts was remarkable – the paper was more than 100 years old. It was one of the highest prices for a diary sold at auction in China.

Family

After Hu's death, Madame Jiang continued to live in Taiwan. She died in 1975; she is buried next to her husband in the tomb facing the Academia Sinica. Son Zu-wang worked as an engineer at China Airlines. He retired in 1980 and died in Washington on March 12, 2005. He is buried next to his parents. Next to their tombs is a four-square-metre stone tablet with this inscription: "This is a tablet in memory of my dead brother Hu Si-du, erected by his brother Zu-wang."

The family never recovered the body of Si-du. Zu-wang's son, Hu Fu, graduated from Cornell University in 1978 – the third generation of the family to study there –and went to work in the U.S. Department of Labour, where he became the chief of the arbitration department. He did not marry.

"A Rare Friend"

The 50-year friendship between Hu and Miss Williams is one of the most remarkable, and moving, themes of this biography. It is indeed astonishing that they were able to sustain their friendship through letters; this is despite living 11,000 kilometres apart, in an era before Internet and cheap telecommunications.

Miss Williams had a stable family and working life in Ithaca, New York. But Hu was constantly moving and had a hectic schedule; despite that, he found the time to write to her. Both remembered the birthday of the other with cards and gifts.

The friendship continued despite Hu's marriage and relationships with other women. Miss Williams would have liked to marry Hu. But she accepted she could not; she did not press him to divorce. She went to great lengths to show respect to Madame Jiang, even though they could not communicate directly to each other. If they had, they would have found little in common. She continued to correspond with Madame Jiang for at least four years after Hu's death.

From her new home in Barbados, Miss Williams made it her mission to preserve his writings. She retyped all the letters which Hu had written to her; she excised passages from some of them, probably of a romantic or intimate nature. She sent the originals to Madame Jiang. She wrote letters to her and to Zu-wang, saying that Hu had been a gift not only to the Chinese people but to all mankind and that he had influenced thousands and thousands of young people.

She also set aside US$5-6,000 for a foundation to translate Hu's works into English. All this was a wonderful act of love and a way to preserve his writings for future generations.

The feelings of each for the other remained intense throughout the five decades. In 1927, he wrote to her: "Throughout these long years I have never forgotten you...I want you, above all to know how much you have given me...Never doubt that our simple friendship will ever be lost. It can't." The next month she replied: "Just a word about us – I shall not write you (and I trust I shall not have in my heart) anything which is not loyal and considerate of your wife – who must really love you a very great deal. It is not disloyal to recognise in you a rare friend, always mentally stimulating, of whom I am very fond. I want nothing else. You are both victims of an unfortunate system ... She perhaps unconsciously, you with unflinching awareness." Miss Williams died in Barbados in 1971, aged 86.

Sources for Chapter Ten

Hu Shih Memorial Hall website and staff

A Pragmatist and his Free Spirit, the Half-Century Romance of Hu Shih and Edith Clifford Williams, Susan Chan Egan and Chou Chih-p'ing, Chinese University Press of Hong Kong, 2009.

"How the Image of Hu Shih changed after 1949", by Dong Ligong, *Yanhuang Chunqiu* magazine, eighth issue of 2013.

"Opening a new mood, but not a Teacher, An Objective Evaluation of Hu Shih's status in Chinese Scholarship", by Zheng Dahua, *Guangming Daily*, April 22, 2015.

Global Times newspaper on auction of Hu's diary, published on 19/10/2020.

Copyright of Photographs

We wish to express our sincere thanks to three institutions for allowing us to use their photographs on the following pages:

Commercial Press of Hong Kong (from the book《胡適及其友人（1904-1948），Hu Shih and His Friends (1904-1948)》 – for photos of the cover page and on pages 31, 34, 45, 70, 90, 95, 107, 111, 112, 117, 131, 146, 153, 171, 175, 179, 193, 199

Cornell University in the United States – for photos on pages 5, 41, 50, 53

The Hu Shih Memorial Hall, Institute of Modern History, Academia Sinica in Taipei – for photos on pages 63, 212, 215, 223, 225

CHINA'S GREAT LIBERAL OF THE 20TH CENTURY – HU SHIH
A PIONEER OF MODERN CHINESE LANGUAGE

Author Mark O'Neill
Editor Donal Scully
Designer Vincent Yiu

Published by Joint Publishing (H.K.) Co., Ltd.
20/F., North Point Industrial Building, 499 King's Road,
North Point, Hong Kong

Printed by Elegance Printing & Book Binding Co., Ltd.
Block A, 4/F., 6 Wing Yip Street, Kwun Tong, Kowloon, Hong Kong

Distributed by SUP Publishing Logistics (HK) Ltd.
16/F., 220-248 Texaco Road, Tsuen Wan, N.T., Hong Kong

First Published in January 2022
ISBN 978-962-04-4918-5

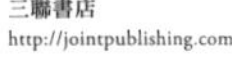

三聯書店
http://jointpublishing.com

JPBooks.Plus
http://jpbooks.plus